Carol Findell

MATHEMATICAL PUZZLES

for Beginners and Enthusiasts

By

Geoffrey Mott-Smith

SECOND REVISED EDITION

DOVER PUBLICATIONS, INC.
NEW YORK

Published in Canada by General Publishing Company, Ltd., 30 Lesmill Road, Don Mills, Toronto, Ontario.
Published in the United Kingdom by Constable and Company, Ltd., 10 Orange Street, London WC 2.

This Dover edition, first published in 1954, is an unabridged and corrected republication of the work first published by The Blakiston Company in 1946.

International Standard Book Number: 0-486-20198-8
Library of Congress Catalog Card Number: 55-3389

Manufactured in the United States of America
Dover Publications, Inc.
180 Varick Street
New York, N.Y. 10014

PREFACE

If you have had fun in solving puzzles, if you like to entertain your friends with puzzles and mental gymnastics, this is your kind of book. Here you will find easy puzzles, hard puzzles, puzzles useful and amusing, puzzles for beginners and puzzles for old-timers, puzzles to challenge your logic, your ingenuity, your knowledge. Some of these puzzles are old favorites; many of them are new ones invented by the author.

The primary object of the book is to entertain. To solve many of the puzzles, you need no knowledge of mathematics other than simple arithmetic. Other puzzles require a knowledge of elementary algebra and plane geometry. Some of the puzzles are solved for the reader in the text, in order to show how to attack more complex puzzles of the same type. Answers to all puzzles are given on pages 139–235, and here the full method of solution is explained for all the more difficult puzzles.

The chapters of the book are arranged in the order in which they should be read by anyone whose schooling in mathematics is not fresh in his mind. The first chapter contains easy arithmetic puzzles, most of which can and should be solved without recourse to pencil and paper. The second chapter takes up puzzles based on logic, to sharpen the reader's ingenuity. The third and fourth chapters present a variety of types of puzzles, ranging from easy to difficult, which can be solved by simple algebra. Geometry is introduced by dissection puzzles, some of which are solved by theoretic considerations and some by simple trial and error. Other aspects of geometry are touched upon in the sixth chapter. The seventh and eighth chapters dig into the properties of digits and integers; they contain the hardest puzzles in the book. Related puzzles of decimation are given a separate chapter following. The tenth and eleventh chapters concern permutations and combinations and probability, a rather specialized field, so that fundamental formulas are given for the guidance of the beginner. The last two chapters analyze some number and board games of a mathematical character and others with elements of mathematical exactitude. While games seem far removed from the formal study of mathematics, they afford an excellent opportunity for the exercise of ingenuity in analysis.

Within each chapter the puzzles are best attacked in the given order, since in some cases a puzzle depends for its solution upon some previous puzzle in the chapter.

In the Appendix are given tables of primes, squares, and so on, together with explanations of how to extract square and cube root. Besides being generally useful to the puzzle addict, these tables are needed for the solution of a few problems in this book.

If the reader wishes to delve further into the theory of mathematical puzzles, he should consult the works of Sam Loyd, H. E. Dudeney, and W. W. Rouse Ball.

Sam Loyd (1847–1910) was a genius in the invention of puzzles of all sorts. Besides being one of the great pioneers in the composition of chess problems, he invented many of the *forms* in which puzzles are now cast. His works were printed mostly in periodicals, but several compendiums of his puzzles have been published.

H. E. Dudeney (1857–1931) was an English mathematician who interested himself in puzzles, and published several collections of his own inventions. He was the first to solve a number of classical problems. The reader of his works must be prepared to find very easy and very difficult puzzles intermixed without warning.

Another English mathematician of the same period, W. W. Rouse Ball, published in 1892 his *Mathematical Recreations,* one of the definitive works on classical problems and the theory of their solution.

The books of these pioneers are out of print, but second-hand copies are fairly easy to obtain, and the books are of course available in many libraries.

Thanks are due to Albert H. Morehead, Rubin Atkin, Lewi Tonks, and L. F. Lafleur for valuable suggestions incorporated in the text, and to my son John for his able assistance in the preparation of the manuscript.

G. M-S.

CONTENTS

Part One

MATHEMATICAL PUZZLES

Part Two

SOLUTIONS

APPENDIX

Part One

MATHEMATICAL PUZZLES

I. Easy Arithmetical Puzzles

1. HOW HIGH IS A POLE? How high is a pole that casts a shadow 21 feet long, if a 6-foot man standing beside the pole casts a shadow 4½ feet long?

How deep is a well, if a rope that just reaches from bottom to top can be wrapped exactly 12 times around the cylindrical drum of a windlass, the drum being 7 inches in diameter?

How many sheep jump over a fence in an hour if 10 sheep jump over a fence in 10 minutes?

2. DOMINO SETS. In a domino set that runs up to double-six, there are 28 bones (pieces). In a set that runs up to double-nine, there are 55 bones.

How many bones are there in a domino set that runs up to double-twelve?

3. MARK-DOWN. A clothing dealer trying to dispose of an overcoat cut in last year's style marked it down from its original price of $30 to $24. Failing to make a sale he reduced the price still further to $19.20. Again he found no takers, so he tried another price reduction and this time sold it. What was the selling price, if the last mark-down was consistent with the others?

4. NINE DOTS. Here is an old puzzle and an easy one; nevertheless, it proves baffling to many a hasty reader. The diagram shows 9 dots in the form of a square. Draw 4 straight lines so as to cross out every dot. You must not cross any dot more than once, nor retrace any line, nor lift the pencil from the paper until all 9 dots have been crossed.

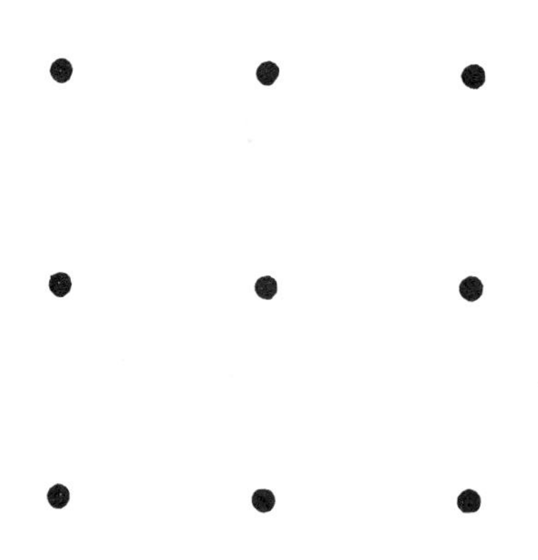

5. MAKING A CHAIN. I have six sections of chain, each consisting of four links. If the cost of cutting open a link is 10 cents, and of welding it together again, 25 cents, how much will it cost me to have the six pieces joined into one chain?

6. THE WILY CHIEF. The following account of conditions on a remote South Sea isle comes from a usually unreliable source. It seems that the M'gmb race inhabiting this isle is ruled over by a wily chief who has a passion for erecting monuments to himself. To do this work he hires men at 5 bmgs per day. But the race is not noted for industriousness, and the chief fines each man 7 bmgs for each working day when he loafs or is absent. Knowing his fellow M'gmbs well, the chief has chosen the rates so that each M'gmb just breaks even in every month of 24 working days. Thus the chief never has to pay out a single bmg. The question arises, how many days does a M'gmb work per month?

7. THE BOOKWORM. The two volumes of Gibbons' "Decline and Fall of the Roman Empire" stand side by side in order on a bookshelf. A bookworm commences at Page 1 of Volume I and bores his way in a straight line to the last page of Volume II. If each cover is 1/8 of an inch thick, and each book without the covers is 2 inches thick, how far does the bookworm travel?

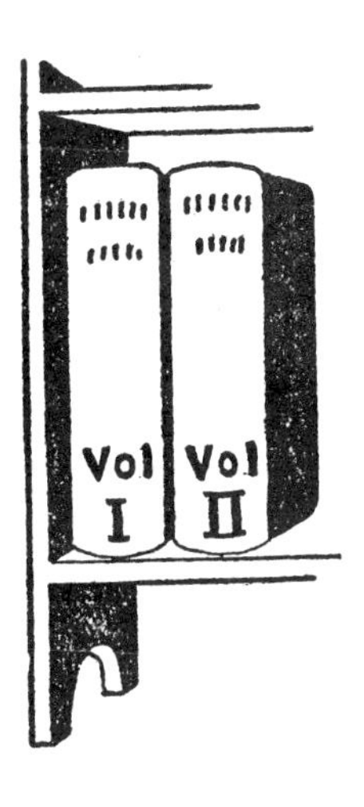

8. AN EASY MAGIC SQUARE. Arrange the digits, from 1 to 9, in a square, so that every row, column, and diagonal totals the same amount.

9. THE FACETIOUS YOUNG MAN. "Give me a pack of Fumeroles, please," said the customer to the young man in the cigar store. "And how much are those Sure-Fire lighters?"

"One Sure-Fire lighter buys three packs of Fumeroles," was the reply.

"Well, give me a lighter. How much is that?"

"The total of the digits of what you owe me is 14," said the facetious young man.

The customer didn't attempt to puzzle that out, but merely gave the clerk a dollar bill and accepted his change.

What is the cost of a Sure-Fire lighter?

10. TANKTOWN TRIOS. Whenever they travel by train, the members of the Tanktown baseball club play pinochle. The nine regulars form three tables of three each. But no outfielder likes to play at the same table as another outfielder, basemen will not sit together, while the pitcher, catcher, and shortstop aver that they see enough of each other on the diamond. Despite these limitations, the nine have been able to organize the three tables in a different arrangement on every trip they have taken. How many different arrangements are possible?

11. WATER, GAS, AND ELECTRICITY. The illustration shows three utility plants, furnishing respectively water, gas, and electricity; together with three houses that are to be serviced. A conduit must be laid from each plant to each house, but it is desired that no two conduits should cross. How can this be done?

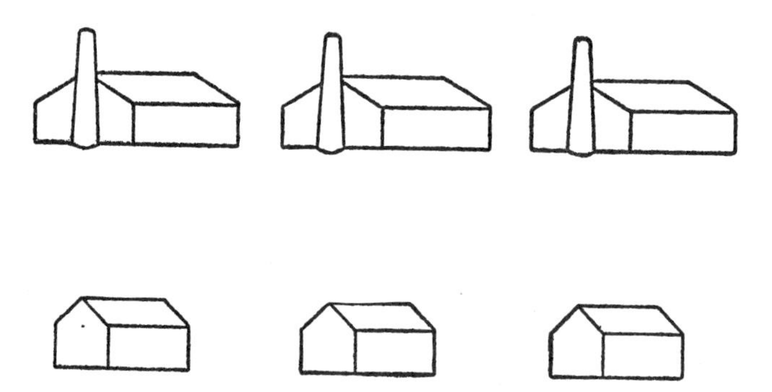

12. A BRICKEY QUICKIE. Anyone who pulls out pencil and paper for this one is disqualified and must go stand in the corner. If a brick balances evenly with three-quarters of a pound and three-quarters of a brick, what is the weight of a whole brick? Quick!

13. SPOTTING THE COUNTERFEIT. "Where is that counterfeit dollar?" the chief of the Secret Service office asked his aide.

"I left it on your desk, along with the eight others that turned out to be genuine."

The chief found the nine "cartwheels" heaped together, with nothing to show which was the spurious coin. He knew that the latter was underweight, so he improvised a balance by setting up a ruler across the lip of an inkwell. He found that by placing coins at equal distances from this fulcrum he could weigh one coin against another with sufficient accuracy to determine whether both were sound dollars.

He then proceeded to spot the counterfeit by just two weighings. This was not a lucky chance; his method assured that no more than two weighings would be necessary. What was the method?

14. THE PAINTED CUBE. A wooden cube is painted black on all faces. It is then cut into 27 equal smaller cubes. How many of the smaller cubes are found to be painted on three faces, two faces, one face, and no face?

15. SHEEP AND GOATS. The illustration shows three sheep (white checkers) and three goats (black checkers), distributed alternately in a line of pens (one row of a checkerboard).

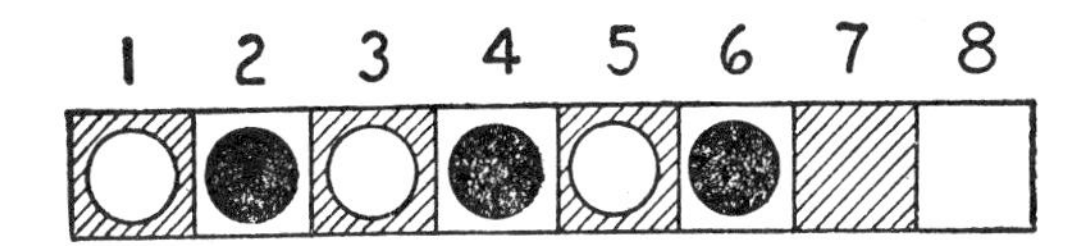

Now we wish to sort out the sheep from the goats, placing the sheep in the pens numbered 1, 2, 3 and the goats in 4, 5, 6. But these are gregarious animals who balk at being moved singly; to move them at all we have to shift a pair of them from adjacent pens to the empty pens. We cannot reverse the order of the pair in transit. For example, we can move 5 into 7 and 6 into 8, but we cannot put 5 into 8 while 6 goes into 7.

After you have found how to sort the sheep from the goats, try to do it in as few moves as possible. It can be done in four moves.

Then, starting from the arrangement shown in the illustration, rearrange the animals so as to get the goats (black) into 1, 2, 3 and the sheep (white) into 4, 5, 6. This puzzle can be solved in four moves.

16. THE BILLIARD HANDICAP. "Do you play billiards? Care to have a game?" asked Huntingdon of the new member at the Town Club.

"Yes, I play," replied McClintock, "but I'm rather a duffer. My friend Chadwick gives me 25 points in 100, and then we play about even."

"Well, I'm perfectly willing to give you a proper handicap. I give Chadwick 20 points in 100. Now let's see—how many points should I give you?"

What is the correct answer, assuming that the stated handicaps are fair?

17. THE SURROGATE'S DILEMMA. "I have come to consult you," said the surrogate to the mathematician, "about William Weston's will. William Weston was fatally injured in a traffic accident while he was on his way to the hospital where his expectant wife was confined. He lived long enough to make a will, which provides that if his child is a boy the estate is to be divided in the proportion of two-thirds to the boy and one-third to the widow. But if the child is a girl, she is to receive only one-fourth and the widow receives the remaining three-quarters.

"Now Mrs. Weston has given birth to twins, a boy and a girl. There is some question whether the will can be held to apply. What would be the correct division of the estate to carry out Weston's evident intentions?"

What was the mathematician's reply?

18. THE LICENSE PLATE. Jim Carter was sorry to have to discard last year's license plate from his car, for the numbers made a beautiful poker hand—a full house. He was disgusted to note that on his new plate all five figures were different. To top that, he inadvertently screwed the plate on his car upside-down, with the result that he increased his registration number by 78,633 until he noticed the error.

What was the number on his license plate?

19. MEASURING TWO GALLONS. "What else can I sell you today?" asked Elmer Johnson, the proprietor of Centreville's general store.

"Well," replied Si Corning, "you'd better give me a couple of gallons of gas. My thrashing machine is a mite low."

"Take five gallons while you can get it, Si. Price is going up, they tell me."

"No, I ain't going to lug five gallons all the way home. Besides, I don't think the tank will take it. Make it two gallons."

"Fact is, Si, I don't have no two-gallon measure. I got an eight-gallon measure, and plenty of five-gallon cans, but I don't see how I can give you just two gallons for certain."

The upshot of the conversation was that Si decided to postpone his purchase of gasoline until he could use five gallons.

But Elmer could have measured out exactly two gallons, using only the 8-gallon and 5-gallon measures. How?

20. MATCHSTICK EQUATIONS. If the after-dinner entertainer were compelled to rely on one article of paraphernalia alone, he could scarcely make a better choice than a box of wooden matches.

The matchsticks lend themselves to the demonstration of feats of equilibrium, of arithmetical and algebraic puzzles, of geometrical puzzles and catches, and to the playing of mathematical games.

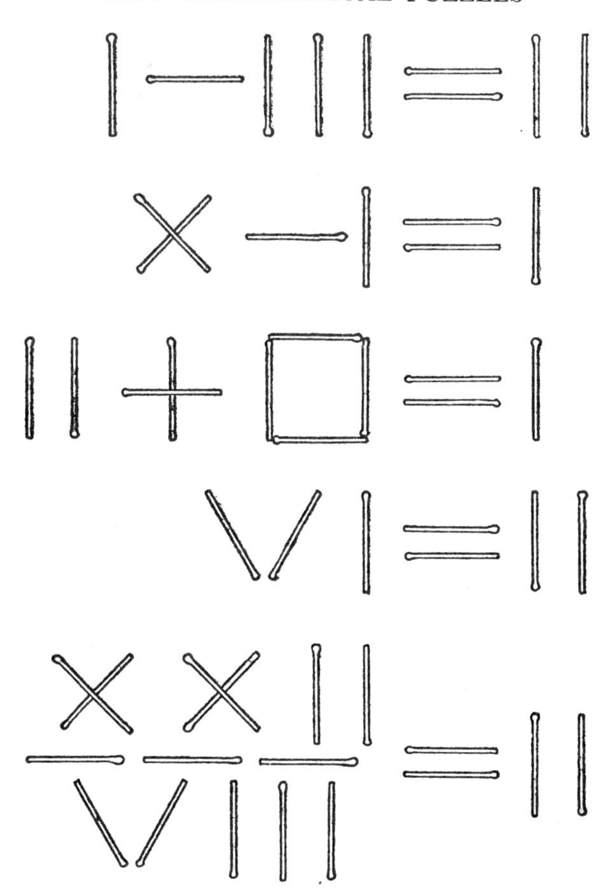

One of the possibilities in this field, little exploited, is the matchstick equation. The illustration gives five examples. Each row is a separate puzzle, an equation given in Roman numerals. In the third puzzle, the square is intended to represent zero, and must be read as such. All the equations are false as they stand, but each can be changed into a true equation by altering the position of only one match.

II. Puzzles of Inference and Interpretation

21. WHAT IS THE NAME OF THE ENGINEER? An oft-quoted problem of the "Caliban" type concerns three pairs of men who shared the names Smith, Robinson, and Jones. The presentation of this puzzle seems to be jinxed; I have heard it misstated numerous times, and in several publications the facts given are either insufficient or contradictory.

Here are the facts as set forth in what may be the original source, the works of H. E. Dudeney.

Three businessmen—Smith, Robinson, and Jones—all live in the "capital district" of New York. (I have changed the locale to the U.S.A. to escape the English currency.) Three railwaymen—also named Smith, Robinson, and Jones—live in the same district. The businessman Robinson and the brakeman live in Albany, the businessman Jones and the fireman live in Schenectady, while the businessman Smith and the engineer live halfway between these two cities. The brakeman's namesake earns $3500 per annum, and the engineer earns exactly one-third of the businessman living nearest him. Finally, the railwayman Smith beats the fireman at billiards.

What is the name of the engineer?

22. AT THE RAINBOW CLUB. Four members of the Rainbow Club sat down one afternoon to play bridge.

In accordance with the rules of the game, they drew cards

from a deck spread face down. The man who drew the highest card chose his seat and the deck to be dealt by his side; second highest took the opposite seat, as his partner; third highest took his choice of the remaining two seats, lowest card becoming his partner.

Without troubling to put the facts in chronological order, we may note that White's card was lower than Brown's. Green asked for a match, which was supplied by White's partner. Black said "What is your choice, partner?" Brown sat on White's left. The left-handed man chose the blue cards, and since Brown is right-handed you can now tell the order of the four players according to the cards they drew.

23. TENNIS AT HILLCREST. Eight men entered the recent tennis tournament at Hillcrest. The tournament was played in three consecutive days, one round per day, and happily no match was defaulted. The first and second round matches were stipulated to be 2 sets out of 3, while the final was 3 sets out of 5. A spectator who was present on all three days reports the following facts:

1. Eggleston never met Haverford.

2. Before play began, Gormley remarked jocularly to Bancroft, "I see that we meet in the finals."

3. Chadwick won a set at love but lost his first match.

4. Altogether 140 games were played, of which the losers won 43.

5. When the pairings were posted, Abercrombie said to Devereaux, "Do you concede, or do you want to play it out?"

6. On the second day, the first-round losers played bridge, and the same table gathered on the third day with Eggleston in place of Abercrombie.

7. Bancroft won 9 games.

8. Franklin won 37 games.

9. The first score of the tournament was a service ace by Gormley, at which Eggleston shouted "Hey, I'm not over there!"

Who won the tournament? Whom did he beat and by what score?

24. WHITE HATS AND BLACK HATS. Three candidates for membership in the Baker Street Irregulars were given the following test of logic. They were told that each would be blindfolded and a hat would be put on his head. The hat might be either black or white. Then the blindfolds would be removed, so that each might see the colors of the hats worn by the other two. Each man who saw a black hat was to raise a hand. The first to infer correctly the color of his own hat would be admitted to membership.

The test was duly carried out. Black hats were put on all three men. The blindfolds were removed, and of course all three raised a hand. Presently one man said "My hat must be black." He was taken into the organization when he proved his assertion to the satisfaction of the judges.

How did he do it?

25. TRUTH AND FALSEHOOD. In a faraway land there dwelt two races. The Ananias were inveterate liars, while the Diogenes were unfailingly veracious. Once upon a time a stranger visited the land, and on meeting a party of three inhabitants inquired to what race they belonged. The first murmured something that the stranger did not catch. The second remarked, "He said he was an Anania." The third said to the second, "You're a liar!" Now the question is, of what race was this third man?

26. WINE AND WATER. Suppose that we have a bucket containing a gallon of water and a demijohn containing a gallon of wine. We measure out a pint of the wine, pour it into the water, and mix thoroughly. Then we measure out a pint of the mixture from the bucket and pour it into the demijohn.

At the end of these strange proceedings, is there more or less water in the demijohn than there is wine in the bucket?

27. FOUR PENNIES. Arrange 4 pennies so that there are two straight lines with 3 pennies on each line.

28. SEVEN PENNIES. Make an enlarged copy of the eight-pointed star shown in the diagram. Place a penny on any point of the star and slide it along a line to another point. Place a second penny on any vacant point and similarly slide it along a line to reach another open point. Continue in the same manner until 7 pennies have been placed on 7 points, leaving only one vacant.

The task sounds easy and *is* easy, but on first attempt the solver usually finds himself blocked after 5 or 6 pennies, unable to place more under the conditions.

29. THE ROSETTE. Now put all the pennies back in your pocket and answer this question without resort to trial.

If we make a rosette of pennies, by putting as many pennies as we can around one penny in the center, so that all the outer coins touch their neighbors and also the center, how many pennies will there be in the rosette?

30. THE MISSING PENNY. This paradox is old, but it is still good. Two market women were selling apples, one at 3 apples for a penny and the other at 2 apples for a penny. (The prices give you some idea of the age of the puzzle!) One day when both were called away they left their stock in charge of a friend. To simplify her reckoning the friend amalgamated the stocks—there were 30 apples of each quality—and sold them all at 5 for twopence. Thus she took in 2 shillings (24 pence).

When it came to dividing the proceeds between the owners, trouble arose. The one who had turned over thirty apples of 3-for-a-penny quality demanded her due 10 pence. The other not unreasonably asked for 15 pence. The sum actually realized was a penny short. Where did it go?

31. THE RUBBER CHECK. A radio dealer was approached by a customer who wanted to purchase a Pandemonium radio, priced at $69.98. The dealer accepted a check for $80.00, giving $10.02 change in cash. Subsequently he endorsed the check to his landlord in part payment of the rent. The check turned out to be worthless and the customer was not to be found. The dealer had to make the check good to his landlord, but the latter accepted a Pandemonium radio in part settlement. As this type of radio cost the dealer $43.75 at wholesale, what was the amount of his loss?

32. MYSTERIOUS COMPUTATION. "Father," said Edward to Professor Digit, "I found this piece of paper on the floor of your study. Do you want to save it?"

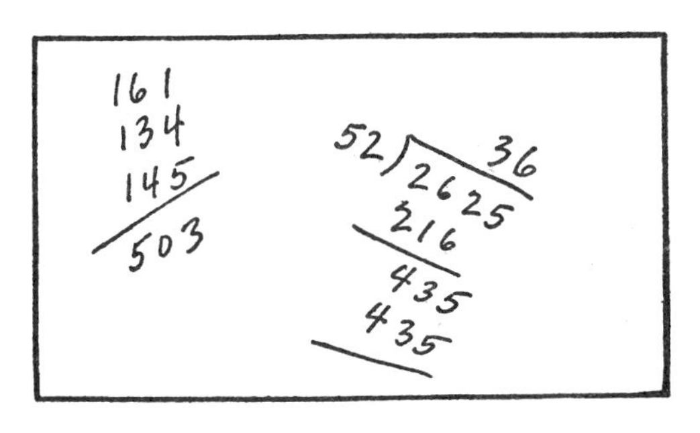

"Let me see. Yes, thank you, I need those figures for a problem I am working on."

"Is that supposed to be an addition, and is that a long division?" asked Edward, pointing to the two groups of figures.

"Yes."

"Well, I guess the teachers couldn't have been very good when you were a boy, because your answers are all wrong."

The professor laughed, and then proceeded to convince Edward that the answers are correct. What did he tell Edward?

33. THE TENNIS TOURNAMENT. If 78 players enter a tournament for a singles championship, how many matches have to be played to determine the winner?

34. TARTAGLIA'S RIDDLE. In ancient times, the neophyte in logic was posed such questions as:

If half of 5 were 3, what would a third of 10 be?

35. STRANGE SILHOUETTES. I have here a familiar object. If I hold a candle under it, the shadow it casts on the ceiling is circular. If I hold the candle due south of it, the shadow it casts on the north wall is square. If I hold the candle due east, the shadow on the west wall is triangular. What is the object?

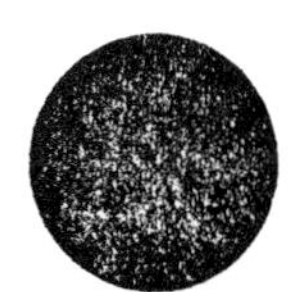  

36. THE DRAFTSMAN'S PUZZLE. Once I propounded *Strange Silhouettes* to a draftsman, and he retorted with a similar puzzle which I think worthwhile passing on.

The diagram shows plan and elevation of a solid block of wood. The broken lines have the conventional meaning that these lines are invisible from the particular angle of sight, but they must be visible from some point, since the block is stipulated to be solid.

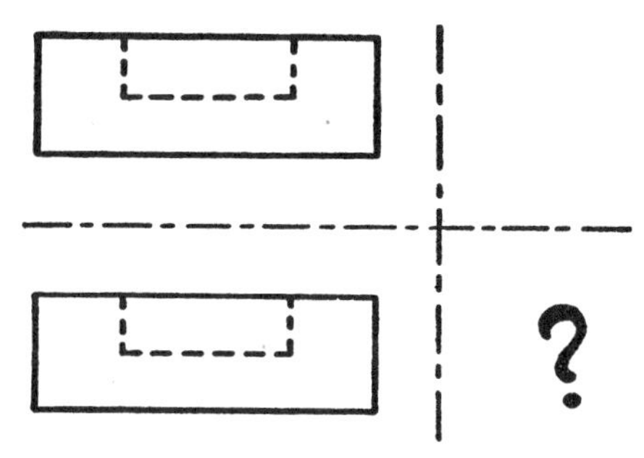

The puzzle is to supply a side view consistent with the other two views. The latter, by the way, are complete; no line, visible or invisible, is omitted for purposes of trickery.

37. A PROBLEM IN PROBABILITIES. If in your bureau drawer are 10 blue socks and 16 grey socks, and you reach into it in the dark, how many socks must you take out to be sure of getting a pair that match?

38. A LAMICED PUZZLE. If you made a business transaction with the Noelomis in the land of Acirema, what would a semid net gain amount to?

39. BEAR FACTS. A bear left its den and went due south in a straight line for one mile. Then it made a 90-degree turn to the left and walked another mile in a straight line. Twice more it made 90-degree turns to the left and walked a mile in a straight line, thus returning to its den. On reaching this starting point, the bear was facing due south.

What was the color of the bear?

40. THE FLAG OF EQUATRIA. The five states that compose the nation of Equatria were once independent principalities. Though having much in common in their outlooks, the people of these states perpetually quarreled with one another, until wise heads resolved upon a union into one nation. Each

principality tried to gain ascendancy through its own claims to superiority, but eventually general agreement was reached on three principles—section, resection, and dissection. To symbolize this credo it was suggested that the triangle be adopted as the emblem of Equatria, and in fact this figure appears on the Great Seal. For ornament, two triangles were combined in the six-pointed star that makes the flag of Equatria so distinctive.

There are five stars in the flag, one for each province. No rearrangement of the stars will ever be necessary, for one of the first acts of the union was to promulgate the Punroe Doctrine, which declares that any attempt by an outside power to gain admittance to the Equatrian union will be resented as contrary to the laws of nature.

As can be seen in the accompanying picture, the stars in the flag are tessellated, each in a different design. An interesting logical exercise is to deduce the plan which governs this tessellation. (No fair asking an Equatrian!)

III. Algebraic Puzzles—Group One

41. THE NATURE OF ALGEBRA. To many persons who have no occasion to use mathematics in later life, the word "algebra" recalls only the memory of certain tiresome scholastic drudgery, a kind of arithmetic where letters are used instead of numbers for the evident purpose of confusion, a frantic pursuit of a mysterious and elusive being known only as x.

But to anyone whose work involves mathematics beyond elementary arithmetic—and this means to virtually any student of abstract or applied science, physical or social—algebra is a wonderful tool, bordering on the miraculous.

Arithmetic and algebra are more than the names of two elementary branches of mathematics. They also indicate two methods of approach and two kinds of objectives that pervade all branches of mathematics, no matter how "advanced." On the one hand, there is a compilation of facts, especially about operations, such as addition, multiplication, factoring, differentiation; on the other hand, there is a constant inquiry into the kinds of classes to which certain facts apply, the characteristics and number of members of each class, the discovery of general propositions from which other facts can be deduced.

Simultaneous Equations

Many algebraic puzzles involve the solution of two or more simultaneous equations.

Example: Three boys picked a number of apples and divided them according to their ages. Edward took three more apples than Wilbur, while David took twice as many as Edward, which gave him eleven more than Wilbur. What was the total number of apples?

If we represent the number of apples taken by each boy by the initial of his name, then the three clauses above tell us that

$$E=W+3$$
$$D=2E$$
$$D=W+11$$

Solving these equations gives $E=8$, $W=5$, and $D=16$, so that the total of apples is 29.

Another method of attack is to use only one unknown. Let x represent the number of apples taken by Edward. Then Wilbur's share is $x-3$, David's share is $2x$, and we are told that $2x=(x-3)+11$. Hence $x=8$, and the other shares may then be computed.

It may sound simpler to use one unknown, where possible, instead of several. It may sound simpler to solve one equation than to solve several simultaneous equations. But all that is an erroneous notion.

If the solver can solve equations at all, a dozen simultaneous equations present no greater difficulty than one. It is not the number of equations that matters; what counts is the specific complexity of the form of the equations. Now, if a puzzle intrinsically involves say five equations, then if all the pertinent facts are put into a single equation instead, this equation is bound to be more complex than any one of the five. And what actually happens when one equation is formed is that the solver performs mentally some of the operations necessary to reduce the five.

To illustrate the point, let us solve Loyd's famous puzzle, "How old is Ann?" The puzzle is stated in a deliberately confusing manner:

"The combined ages of Mary and Ann are 44 years, and Mary is twice as old as Ann was when Mary was half as old as

Ann will be when Ann is three times as old as Mary was when Mary was three times as old as Ann. How old is Ann?"

I suggest that the reader first try the single-equation method. Let x equal Ann's age; adduce from the facts a single equation in x. It can be done—but why do it? How much easier to set up some very simple equations, introducing new literal terms ad lib! Thus:

Let x and y be respectively the ages of Ann and Mary. Then

$$x+y=44 \tag{1}$$

"Mary is twice as old as Ann was . . ." Here is a reference to a past age of Ann. As the definition of this age is complicated, let us at the moment represent it by her present age x less an unknown number of years, a. We are told that

$$y=2(x-a) \tag{2}$$

". . . as Ann was when Mary was . . ." Here is a reference to a past age of Mary, coincident with the time when Ann was $x-a$. At this time Mary's age was therefore $y-a$. Now what is said of this age? ". . . when Mary was half as old as Ann will be . . ." Represent this future age of Ann by $x+b$. We are told that

$$y-a=\frac{x+b}{2} \tag{3}$$

". . . as Ann will be when Ann is three times as old as Mary was . . ." Here is a reference to another and different past age of Mary. Let it be $y-c$. The clause states that

$$x+b=3(y-c) \tag{4}$$

". . . as Mary was when Mary was three times as old as Ann." At the age of $y-c$ Mary was three times as old as Ann. At that time the age of Ann must have been $x-c$. Hence

$$y-c=3(x-c) \tag{5}$$

We have been so prodigal as to use five unknowns. But we have five independent equations, sufficient to find all unknown values, and the equations themselves are all simple in form.

In order to yield unique values for the literal terms, there must be *at least as many independent equations as there are unknowns.* The multiple-unknown method of attack gives an easy check, therefore, on whether the solver has adduced enough facts from the statement of the problem to reach a solution. With many involved puzzles it is easy without such check to overlook a vital fact or two. Note also that the equations (if minimum in number) must all be *independent.* If the simultaneous solution of two or more equations results in an identity, as $3=3$ or $x=x$, then the equations were not all independent.

42. A QUESTION OF BARTER. If the natives of the Weejee Islands rate 2 spears as worth 3 fishhooks and a knife, and will give 25 coconuts for 3 spears, 2 knives, and a fishhook together, how many coconuts will they give for each article separately?

43. SHARING APPLES. A gang of boys made a raid on the Perkins orchard and came back with a quantity of apples, which were then pooled and divided equally among them. Michael said he thought it would be fairer to share by families instead of individuals. As there were two Johnson brothers and two Fairbanks brothers, a redivision by families would have increased each share by 3 apples. With the argument at its height, along came Fred, who, being the oldest of the gang, was appealed to as arbiter. Fred decided that it would be unfair to share by families. Furthermore, he pointed out, he himself would certainly have participated in the raid, to the great increase of the booty, had he not been detained by a compulsory engagement with a rug-beater. But as head of the gang he was entitled to a share. Fred had a way of winning his arguments, so each boy contributed one apple to him, making equal shares all around.

How many apples did the boys gather?

44. A TRANSACTION IN REAL ESTATE. "Jim," said one real estate dealer to another, "I hear you made a pretty piece of change out of that Dingy Street property. They tell me you just sold it for $4,000. I know it cost you only $2,700."

"Your figures are right, but as a matter of fact I lost money," replied Jim in disgust. "There was a big bill for tax arrears hanging over it. I was prepared to pay that, but then I had to put twice as much again into repairs to put the property in shape before I could find a buyer. It was so much trouble that I was glad to get rid of it at a loss amounting to 20% of the taxes plus 10% of the repairs."

Jim didn't specify what the loss was, but the other was able to figure it out.

45. SETTLING THE BILL. After the boxing matches a group of friends went into a restaurant for a midnight snack. "Put it all on one bill," they told the waiter. The bill amounted to $6.00, and the men agreed to split it equally. Then it was discovered that two of their number had slipped away without settling their scores, so that each of the remaining men was assessed 25 cents more. How many men were in the party originally?

46. COWS AND CHICKENS. The same wight who counted sheep by counting the legs and dividing by 4 also kept track of his cows and chickens by counting both the legs and the heads. If he counted 78 legs and 35 heads, how many cows and chickens did he own?

47. THE FARMER'S RETORT. A friend of mine who is a farmer took exception to the answer given to *Cows and Chickens.* Said he,

"How many legs do you suppose there are when a man milks a cow?"

"Why, six."

"Wrong. There are nine."

"How do you make that out?"

"Because the man sits on a three-legged stool."

Now, what answer did the farmer give to *Cows and Chickens* if he counted in at least one three-legged stool?

48. DOLLARS AND CENTS. Sent suddenly on a business trip, George Blake spent half of the money in his pocket on a round-trip railroad ticket and some necessary supplies. Then he bought two newspapers at 3 cents each to read on the train. A taxi from the terminal to his destination took 70 cents. A quarter of what he then had left went for meals and a taxi back to the station. On the return trip he bought a 25-cent magazine. He arrived with as many dollars as at the outset he had had cents, and as many cents as he had had dollars. How much was this?

49. THE JAY ESTATE. Under the will of Jasper Jay, 10% of his estate went to various charities. Son John received 25%, and his share was 25% more than was received by daughter Jill and the baby together. Jill's share was 30 times more than the baby's. After deducting a bequest of $250.00 to Jenkins, the butler, Mrs. Jay as residual legatee received just as much as the two older children together. What was the amount of the Jasper Jay estate?

50. A FISH STORY. Nate Thompson remarked that he had seen a pretty big mackerel and a pretty big pickerel that morning in the shallows of a cove. Pressed for an estimate of their size, Nate observed cautiously that the body of the pickerel was about twice the length of his tail, and about equal to the length of his head plus the tail of the mackerel. The body of the mackerel was about as long as the whole pickerel minus the head. That got us nowhere and we pressed Nate for more details. He "kind of thought" the head of the mackerel was about as long as the tail of the pickerel, while the head of the pickerel was about a quarter of the body of the mackerel. "I jedge," he concluded, "that there was sure three foot of fish there."

It proved to be a pretty big mackerel—if Nate's estimates were correct.

51. WHO NOES? NOT AYE! "If there is no further discussion," said the chairman of the meeting of the Wisteria Improvement Association, "I will put the question to a vote. All those in favor of the motion please stand . . . Thank you. Please be seated. All those opposed please rise . . . The motion is defeated. We will return to a discussion of the original motion, which is to plant a bed of azaleas on the southwest corner of . . ."

"Mr. Chairman!" interrupted a member from the floor. "I thought that was the motion we voted on!"

"No," said the chairman. "We voted on the motion to amend the original motion by substituting the word *begonias* for *azaleas*."

"In that case I would like to change my vote. I misunderstood the question."

From a number of other members came cries of "Me, too!" The chairman read the proposed amendment and called for a new vote. One-third of those who had previously voted *nay* changed their votes to *aye,* while one-quarter of those who had voted *aye* changed to *nay.*

"As matters now stand," said the chairman, "the vote is a tie. I should not like to have to cast the deciding ballot in so important a question. I suggest that the amendment be further discussed."

A member from the Second Ward stood and was recognized. "Mr. Chairman," he said, "some of us have been talking it over, and in the interests of reaching a quick decision are willing to go along with our friends from the Third Ward. We would like to change our vote."

"Mr. Chairman," came another voice, "we have been talking it over too and some of us have changed our minds." The meeting burst into a hubbub, with cries of "Change my vote!" "Let's go along with the Third!" "I much prefer columbine anyhow . . ." After quiet was restored, the chairman said,

"It is evidently the sense of the meeting that we should reopen the whole question. I will therefore once more call for a vote on the amendment to substitute begonias for azaleas."

This time it was found that one-half of the members who had originally voted *aye* and then changed to *nay* had gone back to *aye.* Of those who had changed from *nay* to *aye,* one-quarter went back to *nay.* In addition, one-half of those who had hitherto voted only *aye* decided to change to *nay*. But then one-third of those who had hitherto voted only *nay* changed to *aye*.

"I see," said the chairman, "that the amendment is defeated by a margin of two votes."

How many voters were there at the meeting of the Wisteria Improvement Association?

52. NO FREEZEOUT. Five men sat down to a game of Freezeout Hearts. Each was allowed to buy chips amounting to just 2 dollars. In this game, the loser of each hand pays each other player a number of chips determined by the number of hearts taken in tricks. The first player to lose all his chips is "frozen out" and can no longer play in the game. It was agreed that the player first frozen out should go out for sandwiches and beer while the others continued a four-handed game.

The loser of each hand was also required to pay one chip to the kitty, until it amounted to 2 dollars. The kitty was set aside to pay for the refreshments.

The game went on for several hours, the fortunes favoring none in particular. Long after the kitty was complete, no player had been frozen out.

The players thereupon agreed to play one more round, under "murder or sudden death" rules. In this round, the loser of each hand was to pay each other player a number of chips equal to the number held by that player. In other words, the loser had to double the chips outside his own stack.

The round consisted of five hands, one dealt by each player. Strange to relate, each of the five players lost on his own deal, and when the round was over all players held the same number of chips.

How many chips were held by each player just before the last round?

53. JOHNNY'S INCOME TAX. "Mr. Thompson, will you help me figure my income tax?" asked Johnny, the office boy.

"Sure thing," was the reply. "Bring me your papers."

"Well, here's the form I gotta use, and here's the statement of how much the company paid me during the year."

"Any income from other sources? Odd jobs on the side? Did you take in washing . . ."

"Naw, that's the whole thing."

"Do you claim any deductions? Any capital losses? Any contributions to charity?"

"I gave four bucks to the Red Cross."

"And you have a receipt for it, don't you? Okay, you can claim that as an exemption. I don't suppose you are married, are you? Any dependents? No? Well, then, your personal exemption is $500.00. Your tax is 19% of the taxable net income. I'll work it out for you . . . Here it is."

"Gee," remarked Johnny. "Isn't that funny! The tax is just 10% of what the company paid me. Does it always work out that way?"

"No, indeed," laughed Mr. Thompson. "That's just a happenstance."

What was the amount of Johnny's tax?

54. SPENDING A QUARTER. I purchased some drawing supplies, spending 25 cents for 25 articles. I bought four kinds of articles: paper at two sheets for a cent, pens at a cent apiece, pencils at two for a nickel, and erasers at a nickel each. How many of each kind did I take?

55. THE SPOOL OF THREAD. Mrs. Plyneedle stepped into a dry goods store to purchase a spool of thread. She had in her purse some coins amounting to less than one dollar. She found she could pay for the spool with six coins. On talking over her plans with the salesgirl, Mrs. Plyneedle decided that she had better take two spools at the same price, and found that she could make exact payment with five coins. In the end, she took three spools, and paid with four coins. The salesgirl noticed that had she bought four spools she could have paid with only three coins.

What was the price of the spool, and what coins did Mrs. Plyneedle have in her purse?

56. A DEAL IN CANDY. Three boys received a nickel each to spend on candy. The stock offered by the candy store comprised lollipops at 3 for a cent, chocolate bonbons at 4 for a cent, and jujubes at 5 for a cent. Each boy made a different selection, but each spent his entire 5 cents and returned with just 20 pieces of candy. What were their selections?

57. WHAT SIZE BET? The following incident came to my notice at a poker game in Chicago. The game was stud, with no ante, but with the rule that high hand on the first round of cards face up must make a bet. For lack of chips, the players used coins (none gold).

On one occasion, high hand made a bet comprising 2 coins. Each of the other players stayed without raising. The second hand put 3 coins into the pool. Third hand put in 2 coins and took out one in change. Fourth hand put in 3 coins and took out one in change. Fifth (last) hand put in one coin, then took in change all but 3 of the coins then in the pool.

How much did first hand bet?

58. THE HOSKINS FAMILY. The Hoskins family is a well-regulated household. When it turned out *en masse* to pick blueberries last fall, a separate quota was assigned to the men, women, and children. Each quota was a whole number of quarts, and each individual was expected to harvest exactly as many quarts as every other in his category. The quotas were such that 2 men gathered as many quarts as 3 women and 2 children, while 5 women gathered as many as 3 men and one child. All quotas were filled and the total harvest was 116 quarts. How many men, women, and children are there in the Hoskins family, if there are more women than men and more men than children?

IV. Algebraic Puzzles—Group Two

59. THE TIDE. A motorboat that travels 13½ miles per hour in still water makes a straight run with the tide for an hour and 8 minutes. The return journey against the tide takes 8 minutes longer. What is the average force of the tide?

60. LOCATING THE LOOT. A brown Terraplane car whizzed past the State Police booth, going 80 miles per hour. The trooper on duty phoned an alert to other stations on the road, then set out on his motorcycle in pursuit. He had gone only a short distance when the brown Terraplane hurtled past him, going in the opposite direction. The car was later caught by a road block, and its occupants proved to be a gang of thieves who had just robbed a jewelry store.

Witnesses testified that the thieves had put their plunder in the car when they fled the scene of the crime. But it was no longer in the car when it was caught. Reports on the wild ride showed that the only time the car could have stopped was in doubling back past the State Police booth.

The trooper reported that the point at which the car passed him on its return was just 2 miles from his booth, and that it reached him just 7 minutes after it had first passed his booth. On both occasions it was apparently making its top speed of 80 miles per hour.

The investigators assumed that the car had made a stop and turned around while some members of the gang cached the loot by the roadside, or perhaps at the office of a "fence." In an effort to locate the cache, they assumed that the car had maintained a uniform speed, and allowed 2 minutes as the probable loss of time in bringing the car to a halt, turning it, and regaining full speed.

On this assumption, what was the farthest point from the booth that would have to be covered by the search for the loot?

61. STRIKING AN AVERAGE. A motorist sets out to cover a distance of 10 miles. After he has covered half this distance, he finds that he has averaged only 30 miles per hour. He decides to speed up. At what rate must he travel the rest of the trip in order to average 60 miles per hour for the whole journey?

62. THE SWIMMING POOL. "Hi, Jill, I'll race you to this end of the pool!" shouted Jack, who was then only a few feet from the end he indicated.

"You don't want *much* head start, do you!" retorted Jill, from the other side of the swimming pool. "I'll race you even up for twice the length of the pool."

"Okay," called Jack. "You start there and I'll start here."

They started simultaneously, Jack from the east end and Jill from the west end. They passed each other the first time 20 feet from the east end, and the second time 18 feet from the west end. Who won the race?

That question is too easy. Let's ask another. Assuming that each swimmer maintained his own speed without variation, and turned back instantaneously on completing the first leg of the race, how long was the pool?

63. HANDICAP RACING. On one side of the playground some of the children were holding foot-races, under a supervisor who handicapped each child according to age and size. In one race, she placed the big boy at the starting line, the little boy a few paces in front of the line, and she gave the little girl twice as much headstart over the little boy as he had over the big boy. The big boy won the race nevertheless. He overtook the little boy in 6 seconds, and the little girl 4 seconds later.

Assuming that all three runners maintained a uniform speed, how long did it take the little boy to overtake the little girl?

64. THE PATROL. Immediately the news of the First National Bank robbery was flashed to headquarters, a police car was sent to the High Street bridge. The robbers had made their getaway in a car and the bridge was the only exit from the town on the north side.

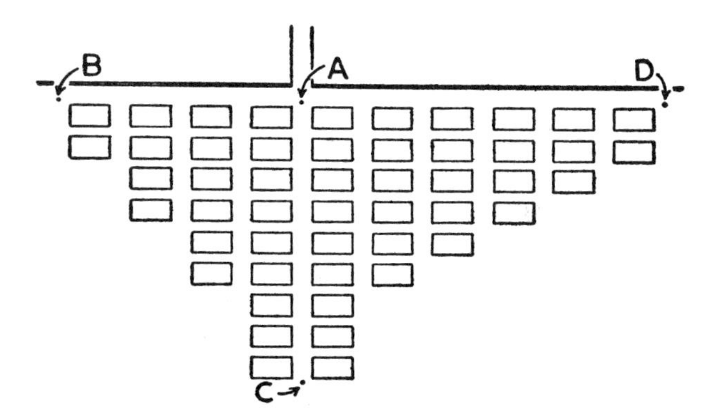

The police car took up a strategic position at point A on the map. Three reserves were instructed to keep watch on the approaches to the bridge. One was assigned to patrol the stretch from A to B; a second was assigned to the stretch A—C, and the third to A—D. Each patrolman walked continuously from one end of his beat to the other, at a uniform rate of 2 miles per hour. All three started from point A at 4:00 A.M.

On the first occasion thereafter when all three met at the police car, a radio message advised the patrol that the robbers had escaped through the south of the town.

If each block in the northern section is 110 by 220 feet

(measured from the centers of adjacent streets), what time did the radio message arrive?

65. THE ESCALATOR. "Henry," said the professor's wife, "you're a mathematician. Tell me how many steps there are in that escalator."

"Well, Martha, they certainly are difficult to count while they are moving. But if you will walk up, and count the number of steps you take from bottom to top, I think we can find the answer. I will start with you, but will walk twice as fast. Just watch me and take one step every time I take two."

When Martha reached the top she reported that she had taken just 21 steps, while Henry had taken 28. The professor was then able to tell her exactly how many steps were in sight at one time on the moving staircase.

66. THE CAMPER AND THE BOTTLE. At 17 minutes past one on a Sunday afternoon a camper embarked in his canoe and commenced to paddle upstream at the rate of 4 miles per hour against a current of 1½ miles per hour. At 5 minutes past two o'clock he drew abreast of a corked bottle floating in the stream. Deciding against stopping to examine it, he continued on his way, only to be overcome presently by curiosity. He turned around, paddled back, and caught up with the bottle just as it reached his camp.

Removing the cork, he found a paper inside, on which was printed in large letters:

HOW FAR DID YOU GET FROM CAMP BEFORE YOU GAVE IN TO YOUR CURIOSITY?

There is no reason why the camper should have paid any attention to this odd message, but you know how these things are. Fortunately he had noticed a large oak tree on the bank just at the point he turned about, so the next day he paced the distance from his camp to the tree and found the answer. Still, don't you think he might have saved himself the walk?

67. HIKE AND HITCH. Fifteen soldiers in charge of a sergeant were detailed to go to a point 60 miles distant. The only transportation available was a jeep, which could carry only 5 men besides the driver. The sergeant undertook to carry the troops to their destination in three loads. As he left with the first party of 5, he ordered the remaining 10 to commence hiking along the road. He unloaded the first party some distance from the goal, with orders to hike the rest of the way. Then he returned until he met the 10, picked up 5 of them, and took them part of the way along the route while the last 5 continued to walk. Finally he returned, picked up the last party, and drove it the rest of the way to the rendezvous. Whether by accident or design—opinions differ—all three parties arrived at the same moment.

The men walked at a uniform rate of 4 miles per hour, while the jeep averaged 40 miles per hour. How much time was saved by the hiking?

68. IF A MAN CAN DO A JOB. "If a man can do a job in one day, how long will it take two men to do the job?"

No book of puzzles, I take it, is complete without such a question. I will not blame the reader in the least if he hastily turns the page, for I, too, was annoyed by "If a man" conundrums in my schooldays. Besides, the answer in the back of the book was always wrong. Everybody knows it will take the two men two days to do the job, because they will talk about women and the weather, they will argue about how the job is to be done, they will negotiate as to which is to do it. In schoolbooks the masons and bricklayers are not men, they are robots.

Strictly on the understanding that I am really talking about robots, I will put it to you:

If a tinker and his helper can refabulate a widget in 2 days, and if the tinker working with the apprentice instead would take 3 days, while the helper and the apprentice would take 6 days to do the job, how long would it take each working alone to refabulate the widget?

69. FINISH THE PICTURE. The picture shows some equalities of weight among objects of four kinds—cylinders, spheres, cones and cubes. At the bottom four cones are placed in the left

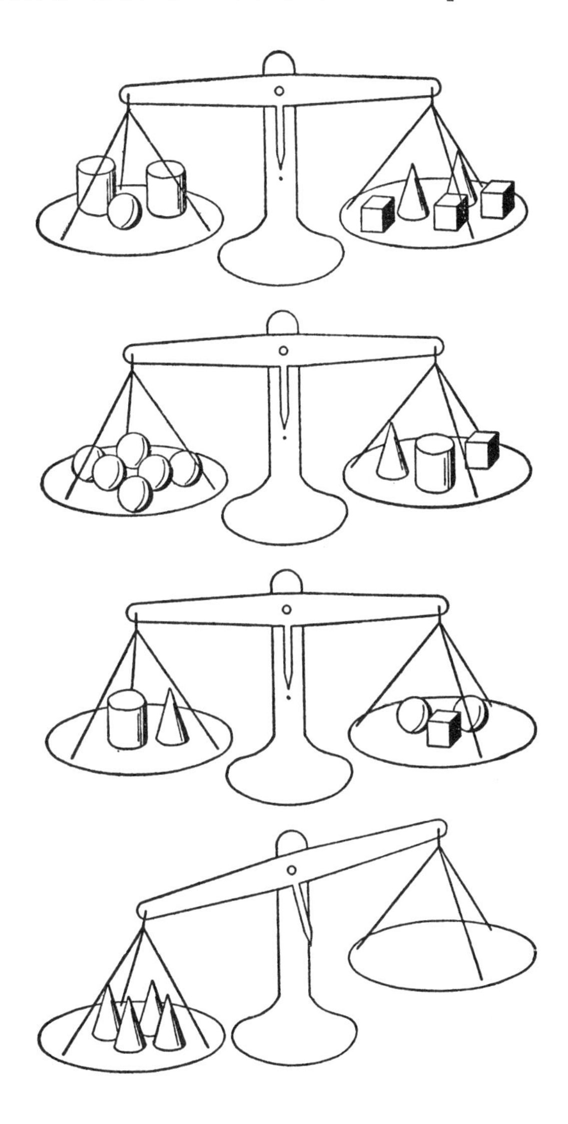

pan of the scales. What is the least number of objects we can put into the right pan to strike a balance?

70. THE ALCAN HIGHWAY. An engineer working on the Alcan Highway was heard to say,

"At the time I said I could finish this section in a week, I expected to get two more bulldozers for the job. If they had left me what machines I had, I'd have been only a day behind schedule. As it is, they've taken away all my machines but one, and I'll be weeks behind schedule!"

How many weeks?

71. SEESAW. Three brothers go to a playground to play on the seesaw. The teeter board has a fixed seat at each end, 5 feet away from the trestle on which the board swings. When Alfred and Bobby take seats, Charles, who weighs 80 pounds, balances them by sitting on Alfred's side 21 inches away from the trestle. When Charles sits in a seat, it takes both his brothers to balance him, Alfred in the other seat and Bobby one foot nearer the center. Now if Bobby takes Alfred's place, where must Alfred sit to balance Charles?

72. A PROBLEM IN COUNTERWEIGHTS. The large flats and other pieces of scenery used in a vaudeville theater are counterweighted by sandbags, so that when they are moved only a small portion of the weight has to be borne by the stagehands.

The theater keeps on hand a set of metal counterweights for occasional use with special pieces. Any or all of the weights can be attached quickly to an elevator rope. There are five weights in the set, so arranged that it is possible to compound any load which is a multiple of 10, from 10 pounds up to the total of all five weights together. The choice of weights is such as to reach the maximum possible total load. What are the several weights?

73. THE APOTHECARY'S WEIGHTS. An apothecary has a set of weights for use in the pans of his scales. He is able by proper selection of weights to measure out every multiple of ½-gram from ½ up to the total of the five weights together. If the arrangement of the weights is such as to reach the maximum possible total, what is it?

Notice that the apothecary can put weights in the same pan as the load he is weighing out.

74. SALLY'S AGE. When Sally went to the polls to vote, the clerk asked her age. "Eighteen," she replied. He looked at her quizzically. "You don't really mean it, do you?" he said. "Of course not!" Sally laughed. "I gave myself the benefit of a year less than a quarter of my real age." The clerk permitted her to cast her ballot, but he is still puzzled as to her true age. Surely you are not.

75. AS OLD AS ABC. Alice is as old as Betty and Christine together. Last year Betty was twice as old as Christine. Two years hence Alice will be twice as old as Christine. What are the ages of the three girls?

76. FUMER FROWNS. When Mr. Fumer returned to his tobacco shop after lunch, he found Joe, his clerk, congratulating himself on a stroke of business.

"While you were out," said Joe, "I managed to get rid of the last two pipes out of that consignment from the Etna Company. A chap came in who wanted a Vesuvius. I showed him that last one we have, and he only wanted to pay a dollar for it. However, we split the difference and he agreed to take it at $1.20. Of course that was a 20% loss for us, but I let it go because he also agreed to take the Popocatepetl pipe at the same price, and that gave us a 20% profit. So we broke even."

Now, why did Mr. Fumer frown?

77. COMPOUND INTEREST. Determine within 5 cents how much I must deposit so that after interest is compounded five times I will have 100 dollars in the bank, the interest rate being 3%.

78. THE SAVINGS ACCOUNT. William Robinson deposited $100.00 in a savings account and left it untouched for 4½ years. At the end of that time (interest having been added eight times) his passbook showed a total of $131.68. What was the rate of interest (within one-half of one percent)?

79. AFTER FIVE O'CLOCK. The clock shown in the illustration has just struck five. A number of things are going to

happen in this next hour, and I am curious to know the exact times.

(a) At what time will the two hands coincide?

(b) At what time will the two hands first stand at right angles to each other?

(c) At one point the hands will stand at an angle of 30 degrees, the minute hand being before the hour hand. Then the former will pass the latter and presently make an angle of 60 degrees on the other side. How much time will elapse between these two events?

80. THE CARELESS JEWELER. On the last occasion I took my watch to a jeweler to be cleaned, he made a careless mistake. He had removed the hands, and in replacing them he put the minute hand on the hour-hand spindle and vice versa. Shortly after I reclaimed the watch I found that the hands were taking impossible positions. But eventually they reached a point where they told the time correctly when read the normal way.

The jeweler had set the hands at 2:00 o'clock. What was the first time thereafter when they showed the correct time?

81. CLOCK SEMAPHORE. (a) At what time between two and three o'clock will the minute hand be as far from VI as the hour hand is from XII?

(b) What is the first time after noon that the minute hand has as far to go to reach XII as the hour hand has passed XII?

V. Dissection of Plane Figures

82. CHANGING A RECTANGLE TO A SQUARE. One of the basic dissection problems is to change a rectangle into a square. The general method is illustrated in the diagram.

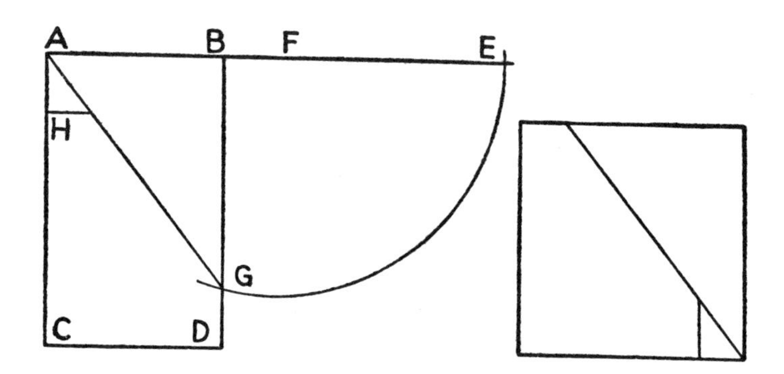

Given the rectangle ABCD, first find the length of side of the square of equal area. Extend AB and measure off BE equal to BD. Bisect AE in F. With F as center and FE as radius strike an arc intersecting BD in G. Then BG is the side of the equivalent square.

Connect AG and cut on this line. On AC measure AH equal to GD. Through H draw a line parallel to CD, and cut on this line.

Slide the triangular piece ABG downward to the right until G lies on CD extended. Then transfer the smaller triangle so that AH coincides with GD. The square thus formed is shown on the right.

83. NOW REVERSE IT. If a rectangle can be dissected into a square, then a square can be dissected into a rectangle. Given the square A, cut it into the minimum number of pieces that can be arranged to form a rectangle one of whose sides is B.

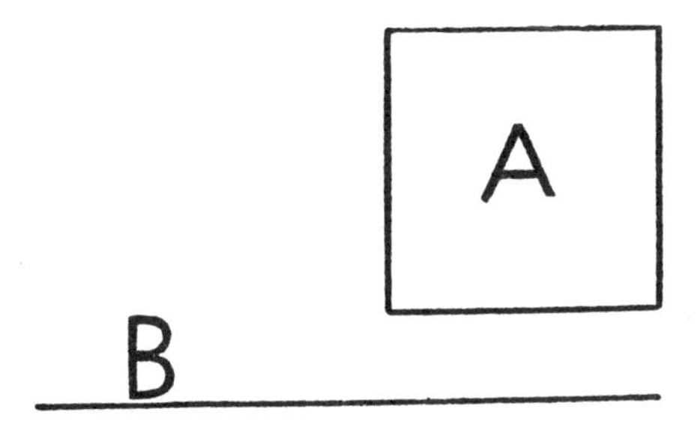

84. THE KITCHEN LINOLEUM. Mr. Houseman wishes to lay down linoleum on the floor of his kitchen, which is exactly 12 feet square. He has a piece of linoleum just sufficient for the purpose, in the form of a rectangle 16 feet by 9 feet. Obviously he will have to cut this piece to make it fit, but he doesn't want to cut it into any more parts than necessary. Fortunately, the linoleum is uniformly brown, without pattern, so that he can cut it in any manner he pleases without spoiling its appearance.

What is the least number of pieces into which the linoleum can be cut to solve Mr. Houseman's problem?

85. THE BROOM CLOSET. Scarcely had Mr. Houseman finished putting the linoleum on the kitchen floor than his wife pointed out that he had forgotten the broom closet. She was most anxious to have the square floor of the closet covered, so Mr. Houseman measured it and later purchased cheaply a remnant

of the shape shown in the diagram, just sufficient in area for the purpose. Mrs. Houseman was aghast at the idea of using checkered linoleum in the closet when the kitchen linoleum was plain brown. But Mr. Houseman remarked that the broom closet was scarcely likely to be exhibited to guests in any event, and he had his way.

"Only mind," said Mrs. Houseman, "that you do not cut that remnant any more than you have to. And don't you spoil the pattern!"

Can you help Mr. Houseman comply with these conditions?

86. SUMMING TWO SQUARES. Given two squares of different size, cut them into the least number of pieces that can be reassembled to form one square.

The squares are assumed to be incommensurable. If they stand in simple integral ratio a more economical dissection may be possible.

87. FROM A TO Z. The report card jubilantly displayed by Tyrus gave his mark in geometry as 100%.

"Humph," remarked his elder brother, Cutler. "I suppose you think you know geometry from A to Z."

"Sure I do."

"Prove it, then." So saying, Cutler drew a large block-letter capital A. "Let's see you divide this A into four parts which can be put together to make a Z. And mind you don't turn any piece over."

Tyrus accepted the challenge and set to work. His first two

attempts were ruled out, because each time he had to turn a piece of the A over to make it fit into the Z. But on the third try he found a solution which Cutler had to admit was correct. Can you find it?

88. THE MITRE FALLACY. Sam Loyd, who invented many ingenious puzzles, once propounded the following:

A carpenter has a mitre, of the shape shown in the diagram —a square with one quarter cut out. He wants to saw the mitre

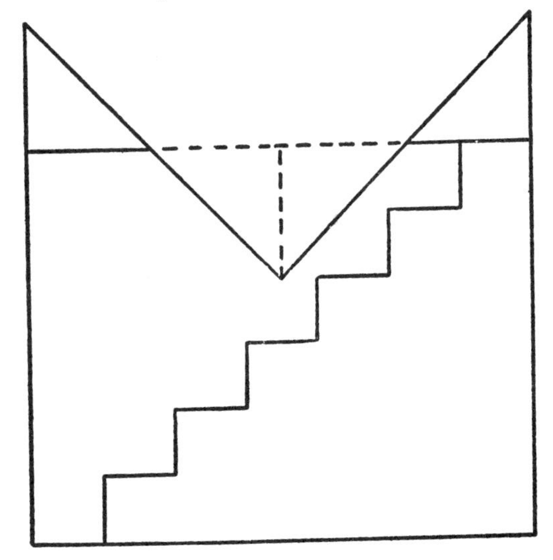

into pieces and fit them together to form a perfect square. What is the least number of pieces necessary and their shape?

Loyd's solution was to cut off the two triangular points as shown, turn them down into the notch to form a rectangle, then cut the rectangle on the "step" principle to make a square. (In this connection, see solution of No. 84—*The Kitchen Linoleum.*)

For once, Loyd slipped. The proposed solution is impossible. Can you prove this statement?

89. WHAT PROPORTIONS? The question here asked will be easier to answer after you have read the solution to No. 88—*The Mitre Fallacy.*

Suppose that a rectangle is capable of being cut into two pieces, in the stepwise manner previously described, which can then be re-arranged to form a square. Suppose that the staircase cut contains thirteen steps in one direction and twelve in the other.

What is the ratio of the width of the rectangle to its depth?

90. THE ODIC FORCE. Reference has been made frequently in literature to a mysterious force known only as "od." Through researches covering the whole period from the cabala of ancient Egyptian astrology to the lexicon of the Twentieth Century crossword puzzle, I am at last able to reveal the precise nature of this force.

It is a property of certain integers and thus has the omnipotence of all mathematics. For example:

$$3\times15=13+15+17=315/7 \text{ and } 1^2+3^2+5^2=7\times5.$$

From these assertions it is readily seen that any square can be dissected into 7 pieces, which can then be arranged to form 3 squares whose areas stand in the ratio 1:3:5.

To discover how to make the dissection you need only use your n-od-dle.

91. THE PIANO LAMP. Here is an easy dissection problem. Lay a sheet of thin paper on the page and trace the outline of the piano lamp. Cut out the silhouette around the lines. Then cut the lamp into pieces which can be re-arranged to form a solid circular disk.

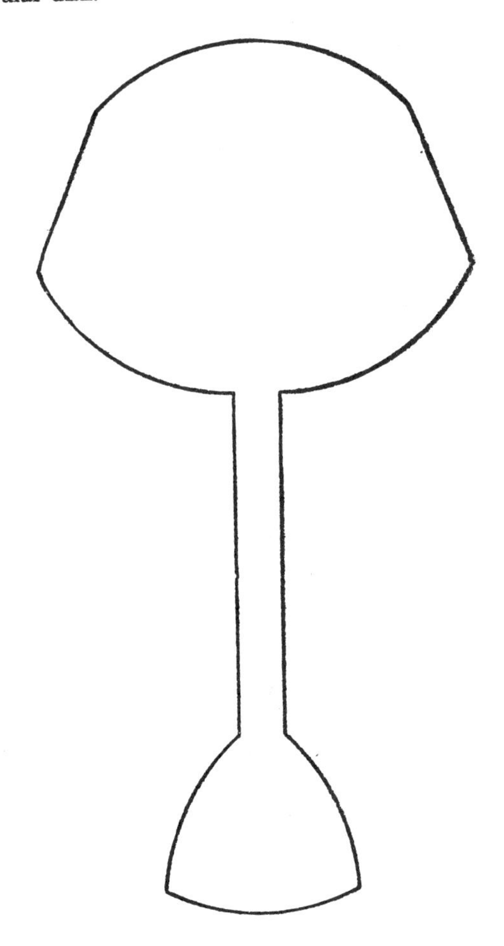

92. CONSTRUCTION OF A PENTAGON. As some of the puzzles in this book involve pentagons, I will here explain how to construct the figure.

Given the circle O in which the pentagon is to be inscribed. Draw two diameters at right angles. Bisect the radius OX in A. With A as center and radius AB lay off AC equal to AB. Then BC is the side of the pentagon.

What is the ratio of the side of the inscribed pentagon to the radius of the circle? This question can be answered without resort to trigonometry.

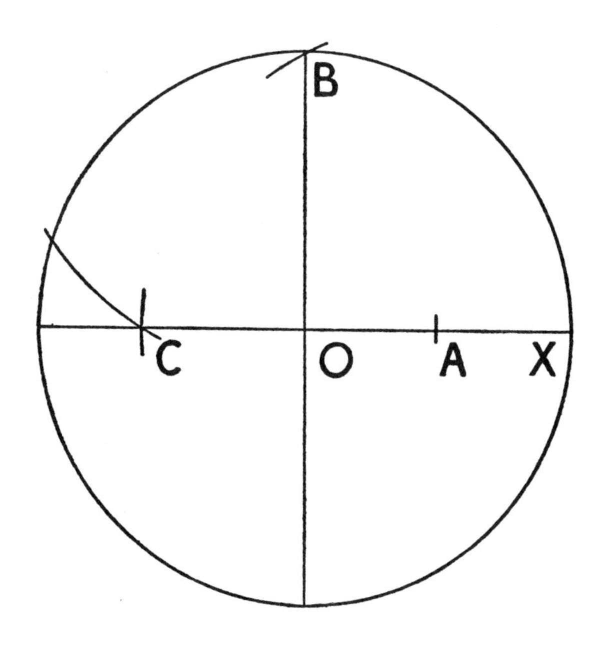

93. THE AMULET. The Pythaclideans, that strange race inhabiting the land of Rectilinea, are said to be very superstitious, as is attested by the fact that no Pythaclid will venture out of his abode without his amulet. This amulet consists of 6 pieces of metal strung together to make a necklace. The number 6 is well-known to have magic properties, being perfect, and it is no wonder that the Pythaclideans deem this amulet to be singularly efficacious in warding off the dreadful Riemanns and Lobatchevskis.

The construction of an amulet involves strange mystic rites. The Geometer or high priest first fashions a square plate of metal appropriate to the suppliant. The absolute size of this plate is determined by the *n*th derivative of his nativity, the submaxillary function of his right ascension, and suchlike matters that do not concern us here. The Geometer then dissects the plate into six pieces which can be rearranged to form a regular pentagon, or into a parallelogram different from the square.

The Pythaclideans fondly believe that they alone are possessed of the secret of this construction, but we believe that it can be rediscovered.

94. FOUR-SQUARE. Here is an easy exercise in dissection. How many different kinds of pieces can be cut from a checkered board, if each piece must contain just four squares and if all cuts must be made along the lines between two squares?

You do not have to cut all the pieces from a single checkerboard 8 x 8. You can have all the board you want for the purpose. We will count as different any two pieces which, although congruent, have the colors arranged in reverse fashion.

95. JACK O' LANTERN. "Oh dear," said Emily, "I can hardly wait until Hallowe'en comes!"

"Why?" said Professor Snippet, her father.

"Because I want to get a pumpkin and make a jack o' lantern."

"Well, we don't have to wait for Hallowe'en for that," he remarked. "I'll make you a jack o' lantern now."

The Professor cut a circular disk out of white paper, then cut the disk into pieces which he arranged on a red blotter to form the jack o' lantern here illustrated. The red, showing

through the holes for eyes, nose, and mouth, gave jack a jolly look that quite captivated Emily. She begged her father to "Do it again!"

Perhaps the reader would like to comply with Emily's request.

96. A REMARKABLE OCCURRENCE. Patrons of the Gambit Chess Club are still talking about the extraordinary behavior of Mr. Sawyer in the matter of his encounter with Mr. Punner.

It seems that the venerable Mr. Sawyer, a charter member of the club, was one evening sitting in its quarters awaiting the appearance of some prospective opponent. Young Mr. Punner chanced to wander in, and Mr. Sawyer asked amiably, "Would you care to play a game?" With the verve of youth the other replied "CHESS!"—thereby intending a pun which he is not the first nor will he be the last to perpetrate.

During his threescore years and ten Mr. Sawyer has no doubt been exposed to this witticism numerous times. At all events, scarcely was the word uttered than Mr. Sawyer jumped to his feet, whipped out a jigsaw, and before the astonished eyes of Mr. Punner proceeded to cut to pieces one of the chessboards. Having rent it to his satisfaction, he then arranged the pieces as shown in the accompanying picture.

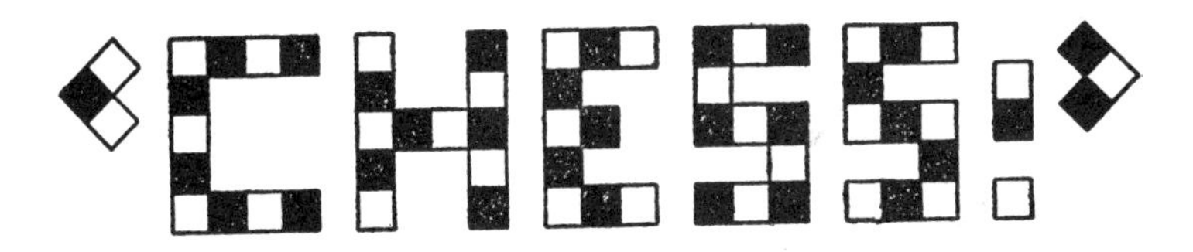

"There!" he snarled at the somewhat intimidated young man. "There, if you please, is the spawn of your side-splitting humor! Take it home with you, nail it on your door, that all passers-by may behold and marvel! Sir, I wish you good evening." Saying which he strode out of the club.

It must be said in behalf of Mr. Punner that he took this reproof with good grace. He was even heard to point out that Mr. Sawyer had performed a rather neat trick, by way of cutting the chessboard into no more pieces than were absolutely necessary in order to form the final tableau.

The reader may be interested to verify this fact.

97. TANGRAM PARADOXES. A long time ago—at least 4,000 years—a Chinese devotee of puzzles dissected a square into 7 pieces, as shown in Fig. 1, and amused himself by arranging the pieces to suggest pictures. From that day to this, interest in the pastime of "tangrams" has never died. In fact, it has acquired a respectable literature, commencing with seven books of tangram pictures compiled in China two millenniums before the Christian

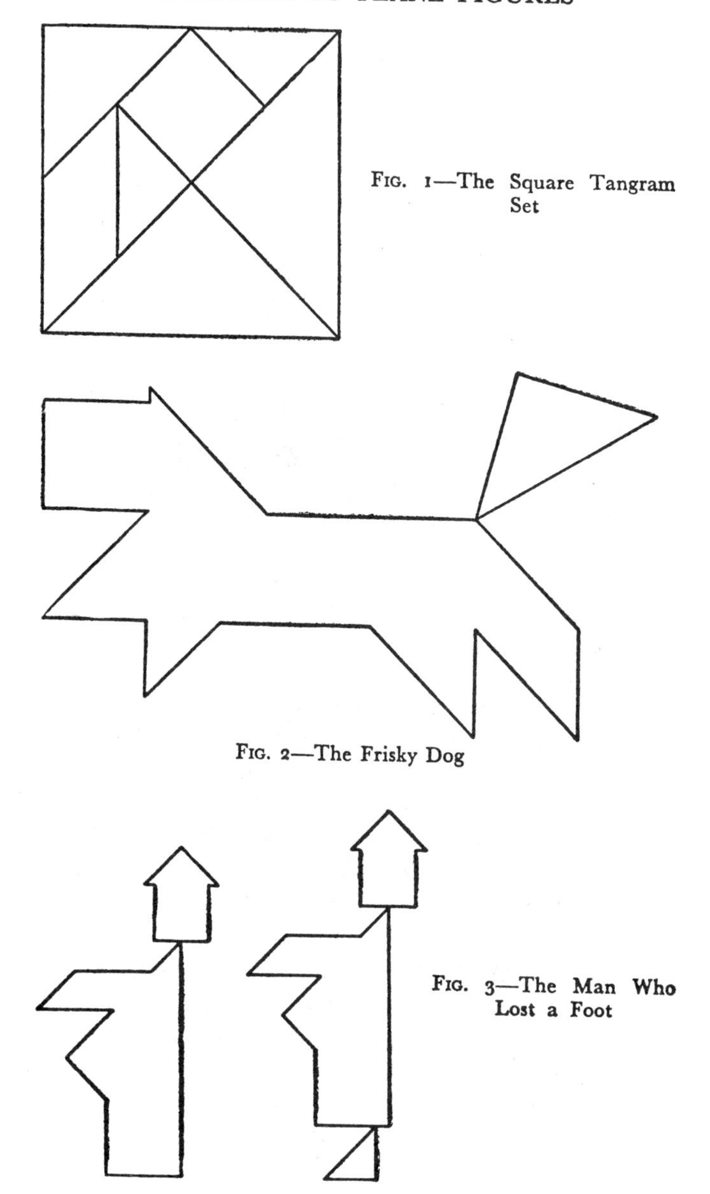

Fig. 1—The Square Tangram Set

Fig. 2—The Frisky Dog

Fig. 3—The Man Who Lost a Foot

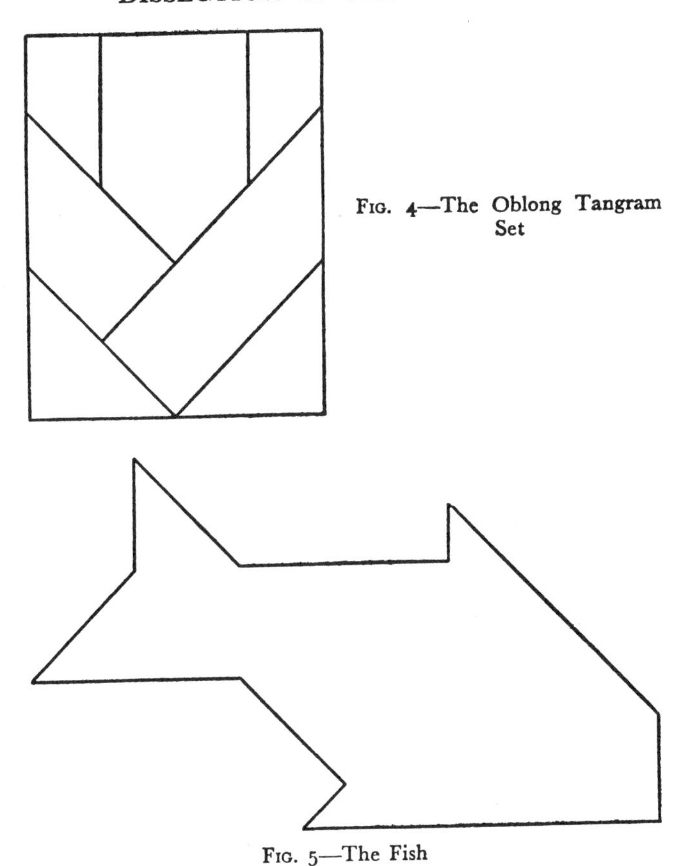

Fig. 4—The Oblong Tangram Set

Fig. 5—The Fish

era, at least two of which have survived. Many other compendiums have been published in modern times.

The invariable practice in making a tangram picture is to use all 7 pieces. The completed silhouette can be regarded as a puzzle: how to form it out of the 7 pieces. Most of the books on tangrams present the pictures in just this fashion.

For example, Fig. 2 shows a frisky dog. The reader will have no difficulty in discovering how to make him.

The tangrams lend themselves to the construction of para-

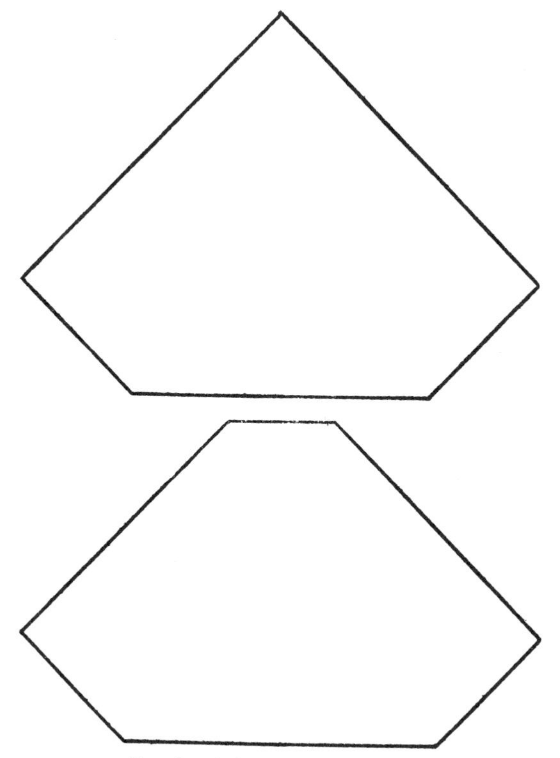

Fig. 6—Grinding the Diamond

doxes, of which Fig. 3 is an example. Here are two men, much alike in appearance, but one has a foot and the other has not. The same 7 pieces were used to construct each figure. Where does the foot come from out of the first figure?

Some years ago, tangrams cut from an oblong, as shown in Fig. 4, achieved wide popularity. I remember a newspaper editorial, commenting in derisive terms on the resurgence of "tangram parties." In other periodicals I found satirical references to the national preoccupation with "how to make the fish." This creature, shown in Fig. 5, was portrayed in the advertisements of a manufacturer of tangram sets.

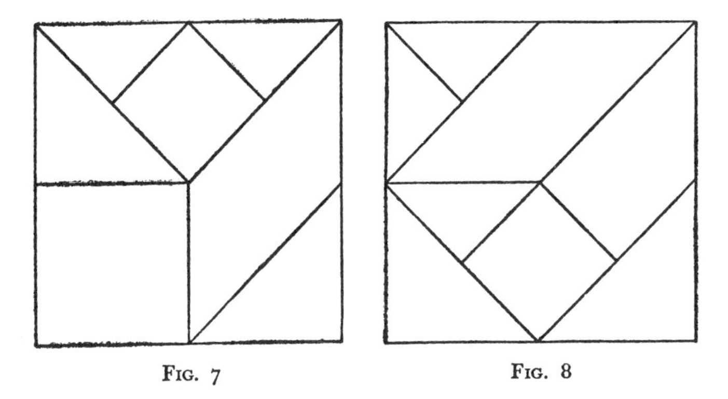

FIG. 7 FIG. 8

Here is a paradox I constructed with the oblong tangram set. Fig. 6 shows the silhouette of a diamond at one stage of its cutting, together with its appearance at a later stage when the point has been truncated. Both figures are made from the same 7 tangrams. What happened to the triangular part at the top of the first picture?

There is of course no limit to the number of different ways a set of tangrams can be cut. The last three diagrams show other sets that have achieved some popularity. Fig. 9 is the only set I have seen which departs from rectilinear forms.

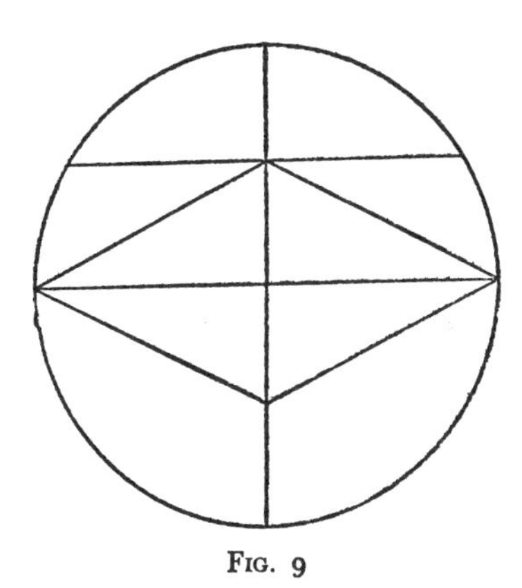

FIG. 9

VI. Geometrical Puzzles

98. THE CLUB INSIGNIA. The Geometry Club of our local high school designed for itself a membership pin in the form shown by the diagram. Archie Mead took the design to Jonathan

Sparks, the jeweler. Mr. Sparks asked, "How large do you want this pin?" Archie replied, "We would like it to be just two-thirds of an inch in diameter. The larger circle of course is the outside edge." "Hum," said Mr. Sparks, "that is going to make the letters GC pretty small. What size do you expect them to be?"

Archie said, "Of course that depends on how much margin is left between the letters and the triangle. I suppose you ought to do whatever you think best on that point. But I can tell you that the side of the triangle will be . . ."

Surely you are as bright as Archie.

99. THE FERRYBOAT GATE. On a recent ferryboat ride, I noticed certain features in the gate across the roadway.

This gate was the usual "lazytongs" affair as shown in the illustration. It consisted essentially of 7 vertical rods, connected by diagonal members pivoted to collars on the rods. The collar at the top of each rod was screwed tightly thereon, but the collars in the lower two tiers were free to slide up and down the rods according as the gate was distended or collapsed.

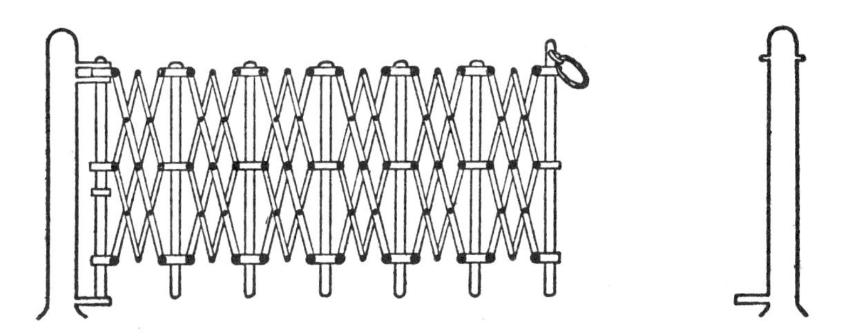

The leftmost rod was pivoted into lugs attached to a heavy post. The rightmost rod could be loosely attached to a similar post on the opposite side of the roadway, by a large ring linked to the rod which could be slung over the top of the post. On the leftmost rod I noticed an extra collar or flange, fixed between the two tiers of sliding collars. The evident purpose of this flange was to prevent the bottom collar from sliding up the rod beyond this point, thus preventing the gate from being distended further.

The puzzle that suggested itself is: How far does the gate reach if pulled out to the maximum distance permitted by the check flange?

I estimated the dimensions of the gate as follows: length of principal diagonal members (between outer pivots), 2½ feet; distance between top and bottom collars (pivot to pivot) on left-most rod when brought as close as the check will permit, 1½ feet; horizontal distance of center of each collar pivot from center of rod, 2 inches.

Assuming my estimates to be exactly correct, how far can the gate be extended?

100. STRIKING A BALANCE. The diagram shows a 60-pound weight on one end of a lever, which has a fulcrum in the

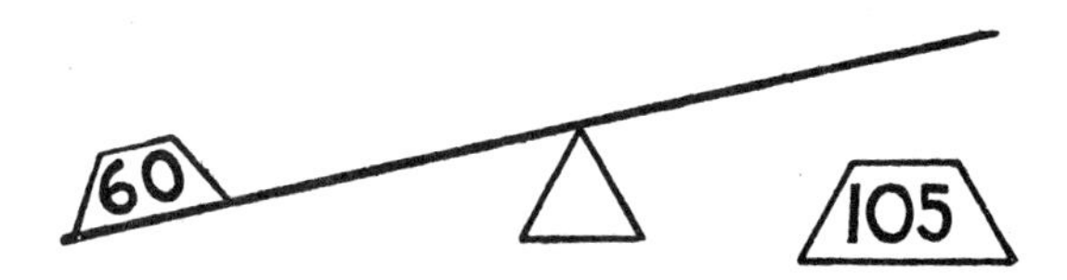

middle. Mark the exact point on the right side of the lever where the 105-pound weight must be placed so that the lever will balance horizontally.

You may neglect the weight of the lever itself. The mass of each weight may be construed to be concentrated at the midpoint of its base.

101. AN INTERCEPT PROBLEM. Three tangent circles of equal radius r are drawn, all centers being on the line OE. From

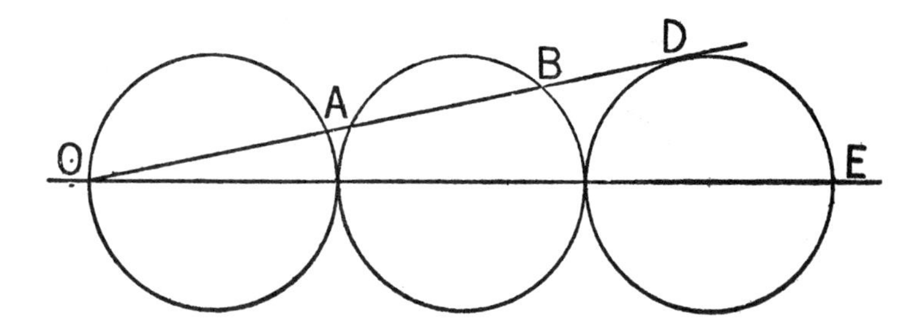

O, the outer intersection of this axis with the left-hand circle, line OD is drawn tangent to the right-hand circle. What is the length, in terms of r, of AB, the segment of this tangent which forms a chord in the middle circle?

102. THE BAY WINDOW. In repainting a house, Mr. Linseed encountered the difficulty shown in the illustration. On one side, a projecting bay window prevented his setting his 20-foot ladder close to the wall. However, he found that he could

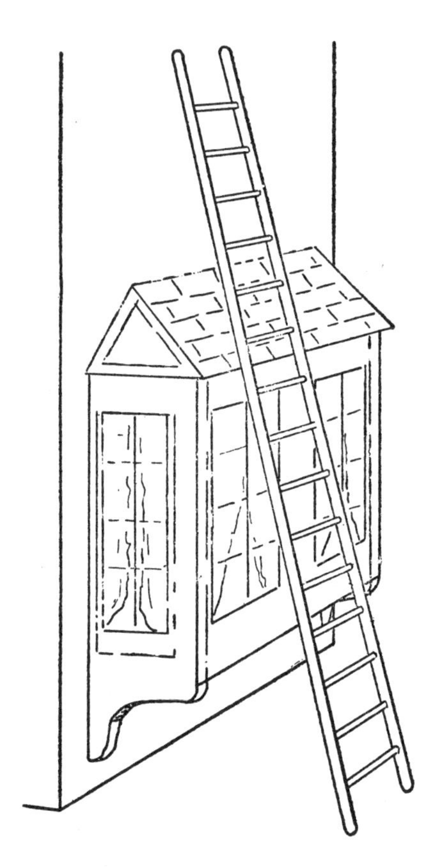

reach some part of the wall above the window by placing his ladder snugly against the bay and also against the wall.

The outer batten of the bay window against which the ladder rests is 3 feet from the wall and 12 feet above the ground. How far up the wall does the ladder reach?

103. THE EXTENSION LADDER. Here is another puzzle about ladders, but it is rather more difficult than *The Bay Window.* In fact, the reader is advised to postpone tackling it until he has read through Chapter VIII, PUZZLES ABOUT INTEGERS.

The accompanying illustration is the scene of the Mayfield Building fire. At this spectacular event, a number of persons were trapped on the roof of the burning structure, and were rescued by the means here depicted. The firemen put up a three-section 70-foot extension ladder in an alley back of the Mayfield Building. The lower end of the ladder was set against the face of the building on the opposite side of the street; the upper end rested against the cornice of the Mayfield Building. A shorter ladder was placed against the latter building at the sidewalk and its upper end was lashed to the extension ladder (at the top of the lowermost section) to brace it. A third ladder was laid nearly horizontally across the street beside this structure, as a traffic barrier. This ladder being a little longer than the width of the alley, one end rested at the base of the Mayfield Building but the other end wedged a few feet up from the sidewalk against the face of the opposite building.

The traffic barrier ladder also was lashed to the extension ladder, and from the fact that a man of average height could just walk under the point where the ladders crossed, without stooping, we may estimate that this point was just 5 feet 10 inches above the sidewalk.

The barrier ladder was 22 feet 11 inches long. The extension ladder projected 2 feet 2 inches above its point of contact with the cornice of the Mayfield Building.

The puzzle is to determine the height of the Mayfield Building above the sidewalk.

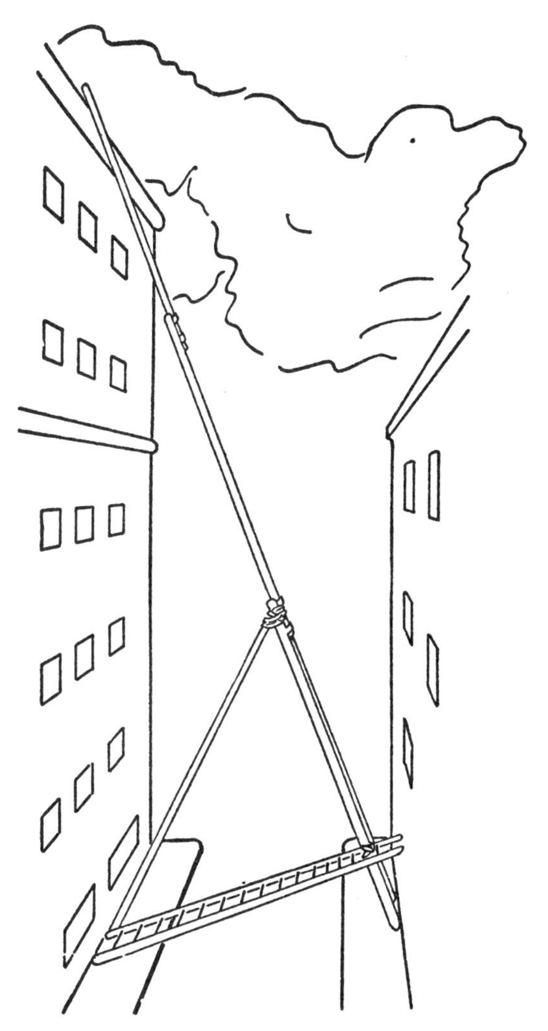

The Extension Ladder
at the Mayfield Building Fire

104. THE SPIDER AND THE FLY. This interesting but well-known puzzle is quoted here by way of preamble to *The Spider's Cousin.*

A spider lived in a rectangular room, 30 feet long by 12 feet wide and 12 feet high. One day the spider perceived a fly in the room. The spider at that time was on one of the end walls, one foot below the ceiling and midway between the two side walls. The fly was on the opposite end wall, also midway between the side walls, and one foot above the floor. The spider cleverly ran by the shortest possible course to the fly, who, paralyzed by fright, suffered himself to be devoured.

The puzzle is: What course did the spider take and how far did he travel? It is understood he must adhere to the walls, etc.; he may not drop through space.

105. THE SPIDER'S COUSIN. It seems that the spider mentioned above had a cousin who lived in the Pentagon Building.

Now this is going to be a very sad story, and anyone whose emotions are easily harrowed had better read no further.

It all happened while the building was under construction. The diagram shows the Pentagon Building in its finished state. Two concentric pentagonal walls enclose a rabbit warren of offices. Access to the offices is gained by 4 circumferential corridors and 8 transverse corridors. At the time of our story, however, only the inner and outer walls and the first floor had been completed.

The spider's cousin was inspecting the structure, with a view to taking space, when one of his spies reported the presence of a particularly succulent fly at the point marked in the diagram—on the inner wall midway between two vertexes, and 9 feet above the floor. The spider, when he received the intelligence, was in the farthest vertex of the outer wall, also 9 feet above the floor.

Our hero immediately set about calculating his shortest course to reach the fly. He would have to go by way of the floor, because the ceiling was not yet in place. That part was easy, but

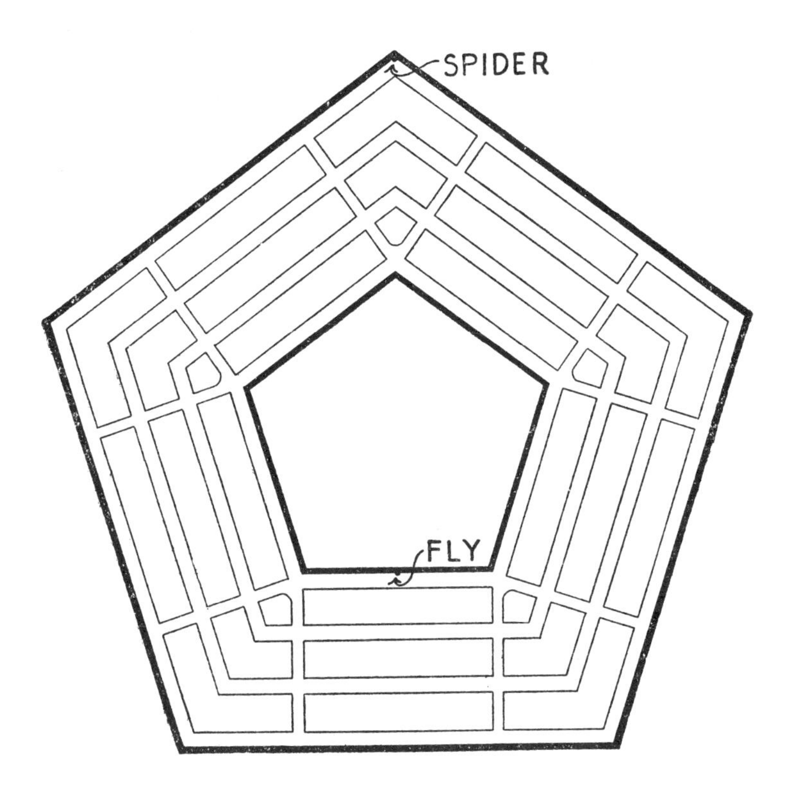

when it came to the angles the spider was baffled. Not having this book with him to explain the construction of a pentagon, he tried to solve the problem by percentage and compound interest, and only got himself in a frightful stew. He rushed off willy-nilly in all directions, and, to make matters worse, found that while he had been cogitating the workmen had put up the interior partitions. Nothing more has been heard from him, and we can only suppose that he is still trying to find his way to any given point.

The arachnid world will be very appreciative of the reader who will solve this spider's problem, toward the day when the Pentagon Building is converted to a riding academy or dance hall.

Oh yes, dimensions. For reasons of security I cannot give the actual dimensions, but the following will serve. Each side of the outer pentagon is 1500 feet; each side of the inner pentagon is 700 feet. These measurements are taken inside the walls, on the planes the spider will have to traverse.

The spiders do not insist on knowing the distance to the fraction of an inch. They will be satisfied with a plan of the route and the distance to the nearest foot.

106. TOURING THE PENTAGON. While we have the illustration of the Pentagon Building before us, let us explore it. Suppose we start from the same point as the spider's cousin. We resolve to traverse every corridor on the floor and return to our starting point. Such a tour is not possible without passing through some corridors more than once, but we want to minimize the number of such duplications. What is the shortest route we can take?

In case you wish to compute the length of the journey, I will mention that the corridors are 12 feet wide, but personally I don't care because I am never going to set out on this marathon!

107. HOW TO DRAW AN ELLIPSE. The illustration shows one way of drawing an ellipse. Fix two pins or thumbtacks on the paper and tie between them a length of string or thread, allowing some slack. With the point of a pencil draw the thread taut, then sweep right around the pins, keeping the pencil point as far away from them as the string will permit.

This construction serves to show just what an ellipse is. It is the *locus* of all points (a *locus* is the path of a moving point that satisfies certain conditions), the sums of whose distances from two fixed points are equal. The fixed points, the pins, are called the *foci* (singular, *focus*). The *major axis* of the ellipse is its width measured on the line of the foci. The *minor axis* is the width measured on a line through the center at right angles to the major axis.

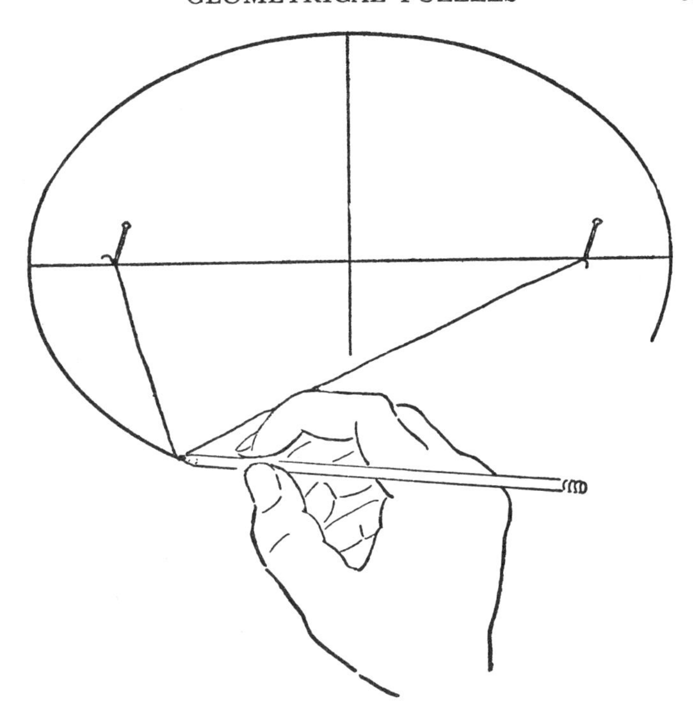

Here is a simple problem based on the construction. If the pins are set 6 inches apart, and the string is 10 inches long, what will be the lengths of the axes?

108. ROADS TO SEDAN. The Allied advance through northern France in August, 1944, was so rapid that forward elements on several occasions outran their maps. The following story is vouched for by an artillery officer of the armored column which, immediately upon the fall of Paris, was sent toward the Belgian border.

This officer was in command of a battery of heavy field guns. Avoiding main highways, the guns were being taken by less-marked routes through forested land in the general direction of Sedan. Reconnoitering planes reported enemy forces in the southern outskirts of the town. Orders came to the battery to halt, prepare for action, and shell the enemy.

The maps then in possession of the commander did not cover the terrain all the way to Sedan. In fact, the battery had almost "stepped off" its last map, a portion of which is here shown. The X marks the position of the battery at the time it received the orders.

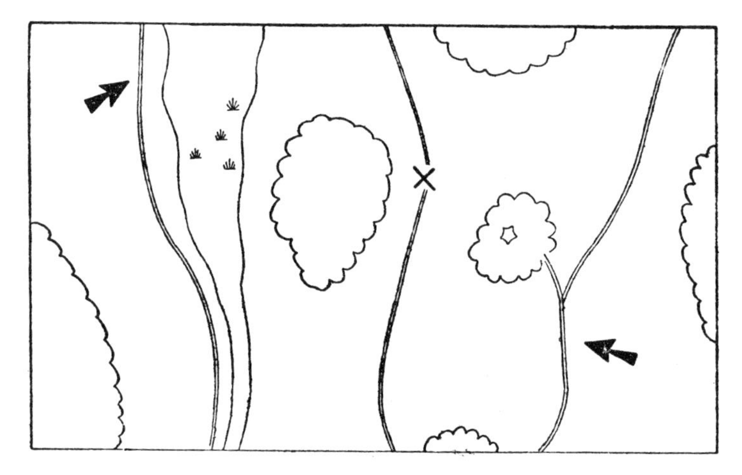

The question naturally arose, how to point the guns upon Sedan, without a map to show the exact location of the town. One of the crew, familiar with the region, was able to state positively that certain portions of the nearby highways headed directly toward Sedan. The portions are indicated by arrows on the map. The suggestion was made that the guns be hauled to one of these roads and aimed thereby.

But the commander was unwilling to delay action or to leave the excellent cover in which they found themselves. Another suggestion was to lay the map on a larger piece of paper, extend two straight lines from the marked roads to an intersection, connect this intersection with the X on the map, and so determine the correct azimuth. The idea was in effect to extend the map so as to plot the position of Sedan. A third suggestion was to orient the map on the ground, sight along the marked portions of the roads to some object on the horizon, then use this object as a target. Both of these suggestions were rejected as impracticable under the circumstances.

The commander solved the problem very simply. He drew a line through X, which, if extended, would pass through the point of intersection of the two road segments (extended). He did not need to find this point of intersection (by using a large piece of paper) to draw his line accurately and so determine the correct azimuth. Finally, he did not even need to use a bow compass—a straight edge was sufficient.

I wonder if the reader can show this construction.

109. THE BILLIARD SHOT. Joe Duffer is playing billiards, and he now finds himself in a quandary. The position is shown in the diagram. The two nearby balls are "red" and "spot"; the isolated ball, "plain," is Joe's cue ball. As you can see, he has an

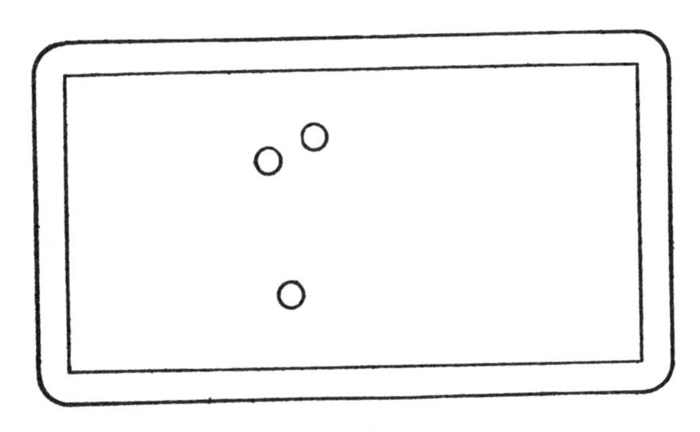

easy carom. Unfortunately, the game is not carom billiards but three-cushion billiards. The rule of this game is that the cue ball must touch three cushions before contacting the second object ball. (No restriction as to when contact with the first object ball must be made.)

Joe might try hitting either red or spot direct, then sending the cue ball around the table to return to the other. But with the object balls so close together, he decides to play a "bank shot." This means that he will send the cue ball around at least three cushions before hitting either object ball, then complete the count by a carom. (Incidentally, the cue ball need not touch three *different* cushions; it may touch the same cushion twice at different times, but such shots are rare.)

Let us advise Joe as to the best three-cushion bank shot to try. We will assume that a ball rebounds from a cushion in a true fashion, the angle of reflection being equal to the angle of incidence. (Actually, the theoretic angle of reflection is usually modified by "English" or "twist" imparted by the player in striking the cue ball or acquired by it through the friction of the cloth.) It is not enough to show Joe the general direction in which to aim; we must show him the exact point at which to touch the first cushion in order to assure the count.

VII. Properties of Digits

110. DIGITS AND INTEGERS. A *digit* is any one of the symbols 1, 2, 3, 4, 5, 6, 7, 8, 9. It is convenient to include 0 (zero) as a digit, although in strict parlance 0 often is excluded.

An *integer* is a whole number, whether expressed by one digit, as 7, or by many digits, as 65,913,448,065,814.

Arithmetic is the science of the fundamental operations—addition and subtraction, multiplication and division. As such it is prerequisite to the study of all higher mathematics, as well as to dealing with the grocer, the banker, the income tax collector. Elementary school arithmetic also makes some study of the *properties* of digits and integers. Of course it just skirts the edge of the subject. Thoroughgoing inquiry into the properties of numbers is left to "higher arithmetic" and "theory of numbers."

The puzzles in this part of the book concern the properties of digits and integers. They venture a little way into the theory of numbers. But I do not expect the reader to be acquainted with the formulas of that fascinating science. On the contrary, my aim is to encourage the reader to rediscover some of these formulas for himself. All the puzzles can be solved by knowledge of arithmetic and elementary algebra, plus a little ingenuity in devising methods of attack.

111. DIGITAL ROOTS. If the sum of the digits of a number is divisible by 3, then the number is divisible by 3. This is only one of the many useful facts that can be inferred from roots.

The *digital root* of an integer is the single integer reached by continued summation of its digits. Given the integer 917,534; to find its digital root: add the digits; the sum is 29; add the digits of 29; the sum is 11; add these digits; the sum is 2: the digital root of 917,534 is 2.

Note that 0 can never be a digital root. There are only 9 possible roots.

All arithmetic operations can be checked by digital roots. The root of the sum, difference, product, or quotient of two integers can be determined by performing the same operation on their roots. For example:

```
    9530624      2
      87235      7
    -------
   47653120      1
  28591872       6
 19061248        4
667143368        5
76284992         2
------------
831803984640     9
```

To check this multiplication, we determine digital roots of multiplicand, multiplier, and supposed product. For the first two we have 2 and 7, whose product is 14, whose root is 5. Therefore the supposed product, whose root is 9, is incorrect. To check the work, we determine the root of each partial product. The first four roots are what they should be; the last is wrong. Here we have 2 instead of 7 ($8 \times 2 = 16$, whose root is 7). Thus we have localized the error.

Digital roots are the basis for the well-known check of "casting out nines." A long column of addition is checked by determining the sum of the roots of the separate additives, and comparing it with the root of the supposed total. In summing digits for this purpose, 9 is subtracted ("cast out") whenever the sum exceeds 9, since the addition or subtraction of 9 leaves the root of any integer unchanged.

Some of the problems in this book involve Diophantine equations. Such an equation is indeterminate in form, but has a unique solution (or a finite number of solutions) through the stipulation that its roots are integral. Discovery of the solution is often easiest (if not necessary) by trial and error, after the field of search has been suitably narrowed. Digital roots often prove a powerful tool to limit the search. The same empiric attack is sometimes simpler than the application of a precise formula. For example:

The sum of a number and its cube is 1,458,275,238. What is the number?

Evidently, the answer can be found by extracting the cube root of the given integer to find the largest integral cube contained in it; the remainder will be the required number. But the extraction of cube root is tedious; let us look for an alternative attack.

We are told that the given integer is of form x^3+x, which can be written $x(x^2+1)$. Then, let us factor the given integer and segregate the factors into two groups whose products are in ratio x and x^2+1. But first we will learn what we can from digital roots.

Possible root of x	1	2	3	4	5	6	7	8	9
Consequent root of x^2	1	4	9	7	7	9	4	1	9
Root of x^2+1	2	5	1	8	8	1	5	2	1
Root of $x(x^2+1)$	2	1	3	5	4	6	8	7	9

The root of 1,458,275,238 is 9. The above table shows us that the required number x must have root 9. That means that it is divisible by 3: consequently x^2+1 *cannot* be divisible by 3. All factors 3 in the given integer will consequently have to be assigned to the group that make up x.

So let us commence factoring by taking out all the 3's. We find that the given integer equals $81 \times 18{,}003{,}398$. Hence x is a multiple of 81.

As there are ten digits in the given integer, x (the largest cube root contained in it) must be a number of four digits, between 1000 and 2000. The remaining factor of x, besides 81, must lie between 13 and 20. Therefore we need examine 18,003,398 only for low factors. Take out 2, leaving 9001699. The lowest remaining factor is 7. Now, we do not have to break down the remaining quotient 1285957; we merely have to be sure that it contains no further low factors. As there are none, we can confidently conclude that $x=81\times2\times7=1134$.

112. THE MISSING DIGIT. If the product of 673,106 and 4,783,205,468 is

3,219,60–,299,743,608

can you supply the missing digit without actually multiplying the numbers?

113. FIND THE SQUARE. One of the following integers, and only one, is a square. Can you find which it is, without actually extracting the square roots?

3,669,517,136,205,224
1,898,732,825,398,318
4,751,006,864,295,101
5,901,643,220,186,100
7,538,062,944,751,882
2,512,339,789,576,516

114. SEND MORE MONEY. What parent of a son in college has never received the following telegram:

```
  S E N D
  M O R E
---------
M O N E Y
```

The odd fact is that this message forms a correct "letter addition." Replace each letter by a digit—the same digit for the same letter throughout, but different digits for different letters—and you will find that the two numbers so formed are correctly totaled below the line.

By way of introduction to the following puzzles of this type, let us solve this one together.

We see at once that M in the total must be 1, since the total of the column SM cannot reach as high as 20. Now if M in this column is replaced by 1, how can we make this column total as much as 10 to provide the 1 carried over to the left below? Only by making S very large: 9 or 8. In either case the letter O must stand for zero: the summation of SM could produce only 10 or 11, but we cannot use 1 for letter O as we have already used it for M.

If letter O is zero, then in column EO we cannot reach a total as high as 10, so that there will be no 1 to carry over from this column to SM. Hence S must positively be 9.

Since the summation EO gives N, and letter O is zero, N must be 1 greater than E and the column NR must total over 10. To put it into an equation:

$$E+1=N$$

From the NR column we can derive the equation:

$$N+R+(+1)=E+10$$

We have to insert the expression $(+1)$ because we don't know yet whether 1 is carried over from column DE. But we do know that 1 has to be carried over from column NR to EO.

Subtract the first equation from the second:

$$R+(+1)=9$$

We cannot let R equal 9, since we already have S equal to 9. Therefore we will have to make R equal to 8; hence we know that 1 has to be carried over from column DE.

Column DE must total at least 12, since Y cannot be 1 or zero. What values can we give D and E to reach this total? We have already used 9 and 8 elsewhere. The only digits left that are high enough are 7, 6 and 7, 5. But remember that one of these has to be E, and N is 1 greater than E. Hence E must be 5, N must be 6, while D is 7. Then Y turns out to be 2, and the puzzle is completely solved.

115. SPELLING ADDITION. Anybody can see that the subjoined sum is correct, but to prove it is another matter. Suppose that you do so by replacing each letter by a digit—the same digit for the same letter throughout. Of course, different letters must be replaced by different digits.

```
  O N E
  T W O
F O U R
-------
S E V E N
```

Any one of several solutions would serve to prove the sum, but for the truism that if your WOES are multiplied (especially fivefold) you are bound to get SORER.

116. ADAM AND EVE. Likely as not, if you order poached eggs on toast in a short-order lunchroom, the counterman will shout to the cook

```
A D A M
  A N D
  E V E
    O N
      A
-------
R A F T
```

What the counterman probably does not know is that this phrase is really an ancient cabala invented by the Numerian astrologers, derived from a sum in addition by replacing each digit by the same letter throughout. Working backwards, you can find several examples of addition that will give this same result, but there is no doubt that the Numerians wished to make the raft as commodious as possible, so that the example they had in mind gives the largest possible total.

117. RESTORING THE FIGURES. When Miss Gates returned to the classroom after recess, she found the janitor just beginning to wash the blackboards.

"Oh, Mr. Benson," she cried, "don't erase that multiplication!" But the janitor had already wiped his wet cloth over a good share of the figures.

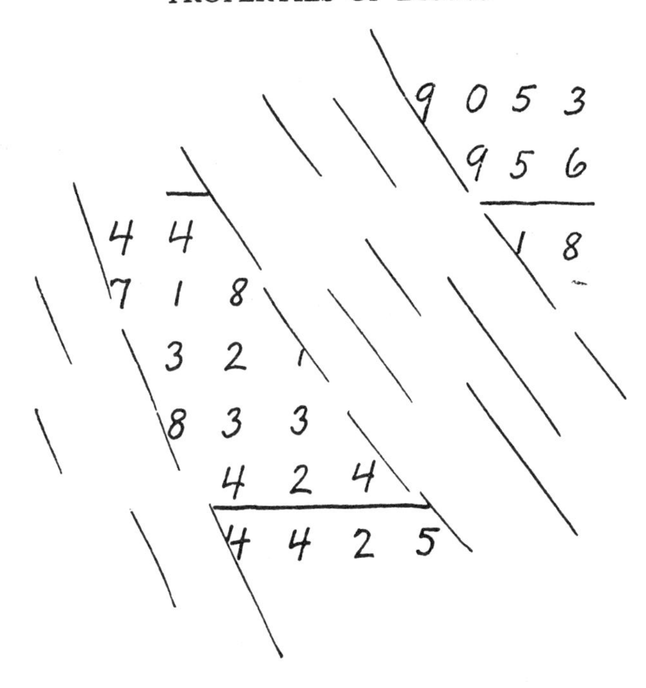

"I'm sorry, Miss Gates, but I didn't know you wanted to save it."

"Well, never mind, it makes a pretty problem this way. I'll ask the pupils to restore the figures."

If seventh-grade pupils can restore the figures—as they did—so can you.

118. LETTER DIVISION. A very popular kind of puzzle, seen many times in magazines, is the "letter division." The working sheet of a long division is presented, with each digit replaced by a letter. (The same digit is replaced by the same letter wherever it occurs.) The puzzle is to "decrypt" the letters by restoring the original digits.

The letter division is an excellent exercise in the simplest properties of digits. Sometimes, unfortunately, this value is nullified by the manner of presentation. The substitute letters are chosen by writing a word or phrase (free of repeated letters) over the digits arranged 1 2 3 4 5 6 7 8 9 0. The puzzle can then be solved by anagramming instead of by arithmetic, and this indeed is the way many persons choose to attack the problem.

Here is a letter division that cannot be solved by anagramming.

```
ABC ) DCEFGA ( FHG
      DGHF
      ----
       AFJG
       AECK
       ----
        DAHA
        DDHH
        ----
          BG
```

119. CRYPTIC DIVISION. In a letter division, the occurrence of the same letter in several places, showing repetitions of the same digit, helps to limit its possible values. And the process of elimination helps, too, e.g., if A equals 7, then no other letter can equal 7.

```
* * ) * 9 * * * ( * * *
        * *
        ---
        * * *
        * * *
        -----
          2 * *
          * * *
          -----
```

Apparently more difficult is the "cryptic division," in which all but a few of the digits are suppressed, and the puzzle is to restore the missing digits with no knowledge of how many different digits are used or where repetitions occur. But the difficulty in fact varies with the particular example. Here is a cryptic division which the solver will find much easier than the letter division given previously.

Replace each star by any digit, so that the whole will be a correct long division.

120. CRYPTIC MULTIPLICATION. Supply the missing digits in this multiplication problem

```
      * * 7 *
        * 7 *
      -------
    * * * * *
  * * * 2 *
8 * 5 *
-----------
* * * * * *
```

121. CRYPTIC SQUARE ROOT. It is astonishing but true that every missing digit in this example of extraction of square root can be correctly inferred, with the aid of only one given 3.

If you have forgotten how to extract square root, see the Appendix. Take note that what we call the *memorandum column* in the examples there given is omitted from the cryptic problem.

```
    *     *     *     *
  -----------------------
 √* *   * *   * *   * *
    *
  ---
    *   * *
        * *
    -------
        * *   * *
        * *   * *
        ---------
          *   * *   * *
          *   * *   3 *
          ---------------
```

122. THE COMPLETE GHOST. It is probably impossible to construct a cryptic division showing not a single digit, and yet provide a unique solution. The record for being the nearest to a complete "ghost" is held by the following puzzle, which presents two related divisions in lieu of any digit.

Replace the stars by numbers so as to make two correct examples of long divisions. It is stipulated that the six-digit quotient of the first example must be the same as the dividend of the second.

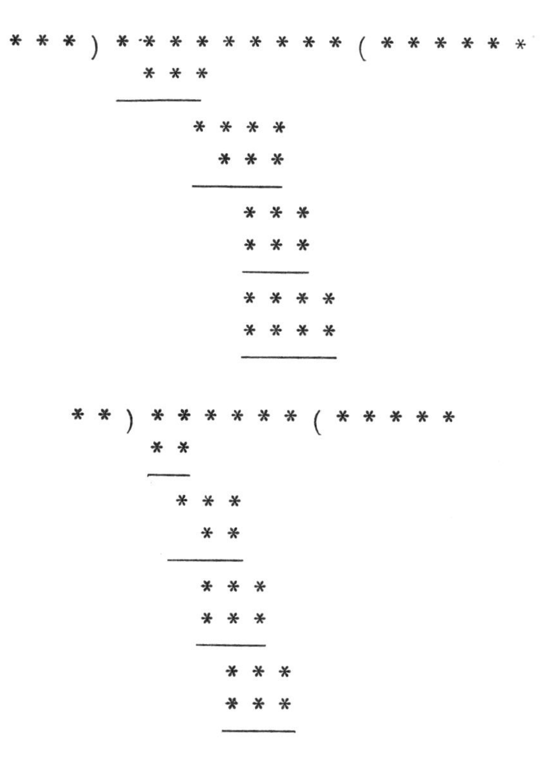

123. A GHOST ADDITION. Here is a puzzle that you could solve entirely by trial and error, but the trick is to use your ingenuity so as to narrow the field of search.

```
    *  *
    *  *
    *  *
    *  *
    *  *
 --------
 *  *  *
```

The problem is to replace the stars in this ghost addition by digits, so that:

(a) The sum is a triangular number not divisible by 3. (For explanation of triangular numbers, see No. 136—*Figurate Numbers.*)

(b) Above the line, each column shows five digits in sequence, picked out of the endless chain 1, 2, 3, 4, 5, 6, 7, 8, 9, 0, 1 . . .

124. NUMBERS AND THEIR DIGITS. In algebra, when we set quantities side by side with no intervening sign, as *abc,* we mean that they are to be multiplied. But when we set digits side by side, as 736, we do not mean the product $7\times3\times6$. Here we express a cardinal number $700+30+6$. If we are told that the digits of a number are, from left to right, $a,b,c,$ we must therefore express the number itself by $100a+10b+c$.

All of the problems under *Two-Digit Numbers* and *Three-Digit Numbers* can be solved by setting up equations of this type.

125. TWO-DIGIT NUMBERS.

(a) What number is twice the product of its digits?

(b) What number is three times the sum of its digits?

(c) What number is the square of its units digit?

(d) What number exceeds its reversal by 20%?

(e) What numbers plus their reversals sum to perfect squares?

126. THREE-DIGIT NUMBERS.

(a) What number is 11 times the sum of its digits?

(b) How many numbers are twice the numbers formed by reversing the order of their digits?

(c) The sum of all three digits of a number is identical with the first two digits, and the sum of the sum is identical with the third. What is the number?

(d) What number is the sum of 17 times its first digit, 34 times its second digit, and 51 times its third digit?

(e) What number is one-fifth of the sum of all other numbers expressed by permutations of the same three digits?

(f) What numbers are the sum of all possible permutations of the three digits taken two at a time?

VIII. Puzzles About Integers

127. PRIME NUMBERS. A number which is not exactly divisible by any other integer (except 1) is called *prime.* A number which can be evenly divided by another is *composite.*

For many purposes in mathematics, it is necessary to be able to determine whether a given integer is prime, and, if it is composite, to determine all of its prime factors.

In elementary arithmetic we learn a few basic tests for factorization: Every even number is divisible by 2; every number whose terminal digit is 5 or 0 is divisible by 5; every number whose digital root is 3, 6, or 9 is divisible by 3; a number is divisible by 11 if the sum of its 1st, 3rd, etc., digits is equal to the sum of the 2nd, 4th, etc., or if the two sums differ by any multiple of 11. There are few simple tests like this for divisibility by higher primes. The task of breaking down certain very large composite numbers has engaged the attention of generations of mathematicians and has led to many discoveries in theory of numbers.

It is possible to solve many problems of factorization by empirical methods. The terminal digit and the digital root of an integer both place a limitation on its possible factors; use this knowledge to narrow the field of search and then find the actual factors by trial divisions.

In the Appendix is given a table of all prime numbers between 1 and 1,000. As a simple exercise I will ask you to determine the next three prime numbers higher than 997.

128. THE SALE ON SHIRTS. "I made a smart move marking down those shirts from $2.00," remarked Mr. Gaberdine to his wife. "We have disposed of the entire lot."

"Good!" said Mrs. Gaberdine. "How much profit did you make?"

"We haven't figured it yet, but the gross from the sale was $603.77."

"Well, how many shirts did you sell?"

Let the reader answer the question.

129. A POWER PROBLEM. The integer 844,596,301 is the 5th power of what number?

130. THE ODD FELLOWS PARADE. The Grand Marshal rode down the street where the Odd Fellows were gathered, instructing them, "Form ranks of three abreast!"

After all had complied, it was found that there was one man left over. As the Marshal did not want this one man to have to march alone, the Marshal changed his order—"Form ranks five abreast!" But after this rearrangement was completed, 2 men were left over.

So the Marshal tried again. "All right men, let's make it seven abreast. Reform ranks!" This maneuver had no more success than the others, 3 men being left over.

"That's what comes of losing so many men to the service!" muttered the Marshal. "Last year we had 497 men in line, so they made even sevens. Well, I'd better try once more . . . boys, we'll try it now eleven abreast."

But with ranks of eleven, it was found that there were 4 men left over. The Marshal decided to waste no more time trying to trim the parade into even ranks, but placed the 4 extra men at the head of the line and gave the order to march.

The question is: How many Odd Fellows marched in the parade?

131. THE UNITY CLUB. It happens that the Unity Club marched in the same parade with the Odd Fellows. In imitation of the latter, they tried forming ranks of 3, 5, 7, and 11, but they had no better luck than the Odd Fellows in coming out even. In ranks of 3 abreast, they had 2 men left over; in 5s, 4 extra; in 7s, 6 extra; and in 11s, 10 extra. What is the least number of marchers there must have been in the Unity Club?

132. CINDERELLA TOASTERS. The president of a chain of retail stores dealing in electrical appliances one day requested the vice-president in charge of sales to produce the figures on the sale of Cinderella Toasters, one of the most popular items sold by the company. The vice-president gave him at once a preliminary memorandum, as follows:

Gross Sales, Cinderella Toasters	
Main Street branch	$3,893.93
All other branches	8,311.19

"Here," said the president to his secretary. "Divide that out for me and let me know how many toasters were sold in the Main Street branch."

"What do the toasters sell for?" asked the secretary.

"You'll have to look that up."

But the secretary found the answer without looking up the price per article.

133. SQUADS AND COMPANIES. The entire standing army of Numeria comprises 1,547 companies of equal size. It could also be grouped into 34,697 even squads. What is the least number of men of which the army can be composed?

134. THE MISREAD CHECK. It was a strange lapse on the part of the bank teller. Evidently he misread the check, for he handed out the amount of the dollars in cents, and the amount of the cents in dollars. When the error was pointed out to him he became flustered, made an absurd arithmetical mistake, and handed out a dollar, a dime, and a cent more. But the depositor declared that he was still short of his due. The teller pulled himself together, doubled the amount he had already given the depositor, and so settled the transaction to the latter's satisfaction.

What was the amount called for by the check?

135. TRANSFERRING DIGITS. The following puzzle, due to Dudeney, is given because the method of solution is useful for a whole class of digital problems.

If we multiply 571,428 by 5 and then divide by 4, we get 714,285, which is the same as the original number with the first digit transferred to the end.

Can you find a number that can be multiplied by 4 and divided by 5 in the same way—by transferring the first digit to the end?

Of course 714,285 would serve if we were allowed to transfer the last digit to the head. But the transfer must be the other way —from beginning to end.

136. FIGURATE NUMBERS. The accompanying diagram shows "Pascal's triangle," which is an orderly way of writing out certain classes of integers called *figurate numbers.*

The top row and leftmost column of this array consist entirely of 1's. Then the table is built up by writing in each cell the sum of the two numbers in the cells at its left and above it.

In the second row (and column) appear all the integers in ascending magnitude; the integers are consequently classed as the *second order* of figurate numbers.

The numbers are called *figurate* because early mathematicians perceived that they give the areas and volumes of certain geometrical figures when built up by discrete units. For example, if you place a number of cannon balls in the form of an equilateral triangle, with n balls in the base, then there will be $n—1$ in the row above it, $n—2$ in the row above that, and so

1	1	1	1	1	1	1	1	1	1	1	1	1	1	1	1
1	2	3	4	5	6	7	8	9	10	11	12	13	14	15	
1	3	6	10	15	21	28	36	45	55	66	78	91	105		
1	4	10	20	35	56	84	120	165	220	286	364	455			
1	5	15	35	70	126	210	330	495	715	1001	1364				
1	6	21	56	126	252	462	792	1287	2002	3003					
1	7	28	84	210	462	924	1716	3003	5005						
1	8	36	120	330	792	1716	3432	6435							
1	9	45	165	495	1287	3003	6435								
1	10	55	220	715	2002	5005									
1	11	66	286	1001	3003										
1	12	78	364	1364											
1	13	91	455												
1	14	105													
1	15														
1															

on until you reach the apex of one ball. The total number of balls in the triangle will be the sum of the integers from 1 up to n. All such numbers are found in the third row of the diagram, and this row is consequently called *triangular numbers.*

Similarly, if you pile cannon balls into a pyramid with a base that is an equilateral triangle, the total number of balls will be some number in the fourth row, which is therefore called *triangular pyramids.*

The numbers that lie along a diagonal line from lower left to upper right, which line is called a *base,* are the coefficients of the terms in a binomial expansion. For example: $(a+b)^4=a^4+4a^3b+6a^2b^2+4ab^3+b^4$. The coefficients 1—4—6—4—1 are seen to lie on the base that starts at the fifth row. Pascal's triangle shows graphically how the binomial theorem is derived, and its relationship to the theory of permutations and combinations.

Many practical problems involve finding the nth term of the rth order of figurate numbers. It is given by the formula

$$\frac{n(n+1)(n+2)\ldots\ldots(n+r-2)}{(r-1)!}$$

In words, the nth term of the rth order is given by the product of successive integers from n to $n+r-2$ inclusive, divided by factorial $(r-1)$.

We see that Pascal's triangle is symmetrical with respect to the diagonal commencing at the upper left corner; consequently the nth term of the rth order is the same as the rth term of the nth order. It does not matter whether we count the orders by rows from top down or by columns from left to right.

For the first four orders, the formula for the nth term reduces to

1st order: 1

2nd order: n

3rd order: $\frac{n(n+1)}{2}$

4th order: $\frac{n(n+1)(n+2)}{6}$

The reader should note especially the formula for a triangular number, as I give a number of puzzles involving triangles.

Just to start the ball rolling, here is an easy question: What is the sum of the first 25 triangular pyramids? Your solution doesn't count if you write out the numbers and add them!

137. LITTLE WILBUR AND THE MARBLES. Little Wilbur has a passion for marbles, and on his last birthday he had signal success in obtaining "miggles" galore from his parents and uncles and aunts.

One day his mother found him seated on the floor laying out his marbles in colorful geometric figures.

"What are you doing, Wilbur?" she inquired.

"I'm playing Chinese Checkers, like you do," was the response. Indeed, Wilbur had arranged his entire stock of marbles into triangles of equal size.

"Well," laughed his mother. "You certainly would need a lot of players for *that* game!"

Then she left, but presently Wilbur called her back.

"See, Mummy, now it doesn't take so many players!" Sure enough, Wilbur had rearranged the marbles into a smaller number of larger triangles, all equal.

"That's very clever, Wilbur. It must have taken a great deal of patience to make the numbers come out right."

"Oh, no, Mummy, it's easy. Look!" And before her astonished eyes the precocious child proceeded to rearrange the marbles four times more, each time making a fewer number of triangles all of the same size.

"I declare," she said, "I never saw anything like it! It must come from your father's side, because I was never able to do the multiplication table."

To cap the climax, Little Wilbur put all the marbles together and made a single triangle.

What is the least number of marbles Little Wilbur could have had?

138. HOKUM, BUNKUM AND FATUITUM. On one of the lesser satellites of Uranus, we are told by juvenile literature, is found a mysterious metal, lighter than aluminum but stronger than steel, with many useful properties such as the power of intercepting gravitation. After intensive study of reports brought back by interplanetary travelers, I have been able to determine the atomic structure of this fascinating element, which is called Hokum. The atomic model is shown on page 86. It is a triangular hexahedron—a solid having six faces, each of which is an equilateral triangle.

While not prepared to give a complete interpretation of this atomic model, I can say that each component represents the sphere of influence of a highly belligerent electron.

The picture shows an atom of power 5—that is, each edge contains five electrons. But the atoms of Hokum are actually of much higher power.

My researches enable me to predict that two new astonishing elements will be discovered in the solar system, if indeed they are not already known. One element is Bunkum, with an atomic power which is one greater than that of Hokum. The other is Fatuitum, whose power exceeds Bunkum by one.

There is good reason to believe that the molecule compounded of one atom each of Hokum, Bunkum, and Fatuitum would have properties almost beyond our experience, such as the ability to square the circle with a red pencil and a piece of string. Such a molecule would have no less than 31,311 electrons, and naturally would be held together only by bonds of the most altruistic nature.

I have been asked to reveal the atomic power of Hokum, Bunkum, and Fatuitum, but will only reply that the answer is obvious.

139. SQUARE NUMBERS. A number which is the product of two equal factors, as 49 (7×7) is called a square. Puzzles involving squares are popular with puzzle makers and solvers alike. The reason is perhaps that the equation $A^n+B^n=C^n$ has an infinity of integral solutions when $n=2$. It is believed that no integral solutions exist if n is greater than 2.

The lowest integers that satisfy the equation are $3^2+4^2=5^2$. The 3:4:5 right triangle has played an important role in history. It was used by the ancient Egyptian "rope-bearers" to lay out right angles and to found the geographical and geometric sciences. Discovery of this triangle is said to have led the Egyptians to the discovery of the very important "Pythagorean theorem"—the sum of the squares on the two legs of a right triangle is equal to the square on the hypotenuse.

It will be noticed that the table of figurate numbers ("Pascal's triangle") does not include the square numbers. How can the squares be derived from this table?

140. SQUARE-TRIANGULAR INTEGERS. The integer 36 equals 6^2 and it is also the sum of the integers from 1 to 8 inclusive; thus 36 is both square and triangular. It is the lowest such integer (above unity). The next lowest is 1225. What are the two next lowest integers that are both square and triangular?

141. PARTITION OF A TRIANGLE. Prove that any triangular number sufficiently large is the sum of a square number and two equal triangular numbers.

142. THE BATTLE OF HASTINGS. There is some question whether the hitherto-accepted description of the Battle of Hastings, given by the Bayeux Tapestry, is correct, in view of the somewhat different version given by the Hooked Rug recently discovered in an attic at Cambric-by-the-Yard. Without entering into this controversy, we may note an interesting circumstance about the latter version.

The Hooked Rug states that the Saxons formed their house carls and knights into a solid square phalanx. The Normans advanced from the shore also in a solid square, but were possessed of "half a thousand more footmenne and full douzaine more of knights." Despite these great odds, the Saxons fought so valiantly that they slew half the foe, losing "only a few score" of their own men, and thus reduced the two armies to exactly the same numbers.

It has been generally overlooked that from this much of the account we can calculate exactly the size of the two armies as they joined battle. I leave the computation to the reader.

143. THE DUTCHMEN'S WIVES. This elegant puzzle dates back at least to 1739. For historical interest, I give it in the original dress, which seems to have imposed the English currency on the Netherlands. I hasten to state that all the American reader needs to know about this currency is that a guinea contains 21 shillings.

Three Dutchmen and their wives go to market, and each individual buys some hogs. Each buys as many hogs as he or she pays in shillings for one hog. Each husband spends altogether 3 more guineas than his wife. The men are named Hendrick, Elas and Cornelius; the women are Gurtrun, Katrun and Anna. Hendrick buys 23 more hogs than Katrun, while Elas buys 11 more than Gurtrun. What is the name of each man's wife?

144. THE CRAZY QUILT. "Well, I declare!" exclaimed Mrs. Thompson. "I thought I had more pieces than that!" She held up a small square of cloth, made by stitching together a number of smaller squares of variegated colors.

"Is that the same quilt you started last summer?" inquired Mrs. Perkins.

"In a way, yes. You see, when the banns were put up for my son Joel, I began to sew him a crazy quilt. It wasn't finished by the time of the wedding, and when Effie came to live with us she was making a quilt too. Her square was larger than mine, but it still wasn't enough for the winter nights. So we decided to put our two quilts together. Our pieces were both the same size—three inches on the side. I ripped the stitches out of my quilt and we added the pieces to hers. We used up all my pieces and then we had one nice big square that was right for the four-poster bed.

"When Joel and Effie built their own house, I urged Effie to take the quilt, but she wouldn't do it. She insisted on giving me back my pieces, because she thought I ought to make myself a quilt too. She said she could get along quite comfortably with the square she had left. These are all the pieces she gave back to me, but I declare! I really thought there were more of them. Not that I care, because I'd be glad to have her take them all. But it does seem strange."

"I think you are right," said Mrs. Perkins. "I remember you showed me your quilt just before the wedding, and I remember that it was big enough to cover that table. Now it doesn't even reach the corners."

What was the size of the crazy quilt displayed by Mrs. Thompson?

145. THE FOUR TRIANGLES PROBLEM. We will commence this fascinating puzzle by outlining a square with matchsticks. The sticks are of uniform length and we will use an integral number of sticks on each side of the square. How large should the square be? That is really the problem!

Then we are going to outline four right triangles, using each side of the square in turn as one side of a triangle. In effect, we will have four right triangles so arranged as to enclose a square space.

All sides of all triangles must be integral multiples of the length of a matchstick. In other words, we must not break any matches to make the figure.

No two triangles may be equal.

What is the minimum number of matches with which such a figure can be constructed?

146. PARTITIONS. By a *partition* of an integer is meant a series of integers of which it is the sum. For example, all possible partitions of 4 are:

4
3 1
2 2
2 1 1
1 1 1 1

How many possible partitions are there of an integer *n*? What is the formula for the number of partitions of an integer into *r* parts? Answers to these questions have as yet not been supplied by mathematical theory. Like certain problems of combinatorial analysis, they seem to defy general solution. If you read what encyclopedias have to say about partitions, you may conclude that the subject consists entirely of the proposition that the number of partitions of an integer into *r* parts is equal to the number of all partitions in which *r* is the largest part.

This proposition, by the way, may be proved as follows. Take for example the partitions of 6 into 3 parts, which are only

4 1 1
3 2 1
2 2 2

Represent the integers in each partition by an appropriate number of points in a column:

4	1	1		3	2	1		2	2	2
.	.	.		.	.	.		.	.	.
.				.	.			.	.	.
.				.						
.										

Now, looking across the rows, you see all possible partitions of 6 in which 3 is the largest part: 3, 1, 1, 1; 3, 2, 1; 3, 3.

The Five-Suit Deck presents a practical problem that actually occurred and had to be solved by empirical methods for lack of theoretical.

147. THE FIVE-SUIT DECK. A few years ago, the proposal was made to include a fifth suit in the standard deck of cards. A quantity of such decks was indeed manufactured and marketed, the fifth suit being variously known as Eagles, Crowns, etc. Contract bridge and other games with the 65-card deck were tried out in many clubs. Anticipating that the new deck might have come to stay, some bridge authorities hastily compiled new tables of chances and rushed into print with revised principles of bidding and play.

I was asked to compute the relative probabilities of all types of hands, classified according to pattern.

By *pattern* is meant the array of integers that sums the number of cards in each suit held in the hand, e.g., 5 4 3 1 is the pattern of a hand that holds five cards of one suit, four of another, and so on.

The table of patterns for the 52-card deck was computed many years ago and has been in constant use for the deduction of strategical principles. The total number of patterns is 39, of which the five most frequent are

4 4 3 2
5 3 3 2
5 4 3 1
5 4 2 2
4 3 3 3

As a preliminary to computing the frequencies of various hands with the five-suit deck, it was necessary to write out all the possible patterns.

An orderly way to do this is to start with the greatest possible concentration of cards into a few suits and proceed toward the more equal distributions, thus

13 3 0 0 0
13 2 1 0 0
13 1 1 1 0
12 4 0 0 0

(In five-suit bridge, each hand was dealt 16 cards, and the 65th card was used as a widow.)

Having written out the patterns in this manner, I looked for some check upon their total number. These patterns are of course the partitions of 16 into any number of parts not in excess of 5, and with the proviso that no part may exceed 13 (since there are only 13 cards in each suit). But on looking into the scant literature on partitions, I found that theory could not predict how many such partitions there should be.

It is evident that the count of partitions of an integer n is compounded from the count of partitions of lesser integers $n—1$, $n—2$ . . . 2, 1. I therefore made a table of the count of partitions of integers 1, 2, 3, up to 16. The precaution was well-advised, for I found that in writing the patterns "by eye" I had overlooked two of them.

I found that simple rules could be formulated as to how to build up the table of partitions so as to avoid error. What I propose to the reader is that he shall rediscover these rules by making a table up to, say, the integer 8.

148. ORDERS OF INFINITY. When we say that a group of objects is countable or *enumerable* we mean that its members can be arranged in one-to-one correspondence with the integers 1, 2, 3, etc. This series is infinite in extent—we can always write an integer larger than any given integer—but it is *discrete*. Between any two adjacent members of the integer set, as 7 and 8, lies a void, containing no other members of the set.

An example of a countable set is the prime numbers. Although infinite in extent, the prime numbers are discrete. Between 31 and 37 lie no other primes.

There is a higher order of infinity than is possessed by the integers. The class of all real numbers, both rational and irrational, is not enumerable, for the reason that it is not discrete. In whatever way the numbers are arranged, it will be found that between any two adjacent members lie an infinitude of other members of the set.

By a *rational* number is meant one that can be expressed as the quotient of two integers. Thus a number may be rational although it is a never-ending decimal, e.g., $5/7 = .714285714285\ldots$ An endless decimal, if rational, sooner or later reaches a group of digits which it thereafter repeats forever.

An *irrational* number cannot be expressed as the quotient of two integers, but only as the sum or limit of an infinite series, and the computation of its value produces an endless non-repeating decimal. A familiar irrational number is π, $3.14159\ldots$

A class which is not discrete is called *continuous,* and the kind of infinity possessed by the real numbers—external and internal, as it were—is called *the continuum.*

Now, viewing the rational numbers apart from the irrational, it may seem that the former in themselves are a continuous class. For example, between $2/3$ and $3/4$ we can find an infinity of other rational numbers of intermediate magnitude, e.g., $17/24$.

But the rational numbers are in fact enumerable. The task here posed the reader is to prove this fact, by arranging them in a demonstrably discrete order.

IX. Decimation Puzzles

149. TURKS AND CHRISTIANS. In former times, the lawful penalty for mutiny was to execute one-tenth of the crew. Custom ordained that the victims be selected by counting off every tenth man from a random arrangement of the crew in a circle. Hence the term *decimation*. In the course of time this term has come to be applied to any depletion of an assemblage by a fixed interval, regardless of whether this interval is ten or another number.

A very old puzzle about decimation is *Turks and Christians*. The story goes that 15 Turks and 15 Christians were aboard a ship caught in a severe storm. The captain decided to propitiate the elements by throwing half his passengers overboard. In order to leave the selection of victims to chance he arranged all 30 in a circle, with the announced intention of counting off every thirteenth man. A clever Christian pointed out to his fellows how to take places in the circle so that only the 15 Turks would be counted out for jettison. What were these places?

150. BOYS AND GIRLS. Five boys and five girls found five pennies. A dispute over ownership ensued, and it was decided to arrange the group in a circle, count out individuals by a fixed interval, and give each a penny as he or she left the circle. The plan was advanced by a clever but unscrupulous boy, who so arranged the circle that by counting a certain girl as "one" he could count out all the boys first. But the girl counted as "one" insisted on her right to choose the interval of decimation, and her astute choice counted out all the girls first.

The arrangement of the circle was thus:

1	2	3	4	5	6	7	8	9	10
G	G	G	B	B	B	G	B	B	G

The count in each case starts with the girl at the extreme left, goes to the right, and then returns to the leftmost individual remaining. Each individual counted out steps out of the circle and of course is not included in the count thereafter.

What interval will count out the five boys first, and what will count out the five girls first?

151. NATIVES AND BRITONS. A party of explorers in Africa, five Britons and five native porters, fell into the hands of a savage chief, who consented to release them only on condition that half the party submit to flogging. The ten men were arranged in a circle in this order:

1	2	3	4	5	6	7	8	9	10
B	N	B	N	N	B	N	B	N	B

The five men to be flogged were to be selected by counting around the circle by a fixed interval until five were counted out. The Britons had arranged matters so that they could first count out the five natives. But the chief did the counting and chose such a number as to select the five Britons.

What interval and what starting point did the Britons and the chief respectively have in mind? (The count goes only in one direction around the circle, being indicated in the diagram by ascending numbers.)

152. JACK AND JILL. It is not generally known that Jack and Jill had some brothers and sisters, and when the question arose who was to fetch the water they decided to settle the matter by lot. The five girls arranged themselves in one circle and the four boys in another. Somebody named a number at random, and each circle was decimated clockwise by this number until only one boy and one girl were left. Jack counted for the boys, beginning with his left-hand neighbor as "one," and Jill counted for the girls, also beginning at her immediate left. Well, you know what happened; it seemed pretty hard lines to Jack and Jill that *that* particular interval of decimation was chosen—what was it, by the way?

153. OUT AND UNDER. When I was a boy, one of my pals showed me a "trick" he had just learned. From a deck of cards he took the 4 kings and 4 queens. He arranged the 8 cards out of my sight, then proceeded to deal them out thus:

The top card was placed face up on the table. The second was transferred from top to bottom. The third was thrown out face up, the fourth transferred to the bottom, and so on. The cards as they were thrown out face up came in the order, kings and queens alternately.

My playmate challenged me to discover the correct initial order of the cards to produce this result. By trial and error I discovered the secret. Later, on thinking it over I found an easier way to the solution. The next time we met, I handed him a full deck of 52 and asked him to deal it "out and under." When he did so he was astonished to see all four suits come out in sequence.

I will not ask the reader to reconstruct the prearranged order for the 52 cards, but I am curious to know how quickly he can arrange the 13 cards of one suit so that they will deal "out and under" in order: A, 2, 3 . . . 10, J, Q, K.

154. THE NIGHTMARE. Following the incidents narrated in *Out and Under*, my chum had a nightmare. He dreamed that a demon perched on his bed, produced a monstrous deck of cards, and commenced to deal "out and under." The cards were evidently not the usual kind, being simply numbered from 1 up. It was evident that the deck had been arranged in ascending sequence from the top down.

"Now," said the demon, "I shall abolish all candy stores and all comic magazines if you do not instantly tell me what card will be the last to come out. Also, you must tell me when card No. 288 will come out, and what number will be on the 643rd card I turn up."

Terrified and dismayed, the boy struggled to speak in vain; the demon dissolved in a burst of light, and the boy found himself alone with the morning sun streaming through the windows. He jumped into his clothes, toured the neighborhood, and was comforted to discover that the demon had not yet carried out his threat.

On hearing my chum's tale I remarked that we should be able to find the answers to the riddles if we but knew the number of cards in the demon's deck.

"Well," said he, "I did notice that the bottom card was No. 971."

Forthwith we took out insurance against a major catastrophe by figuring out the answers to the demon's questions. I should be glad to tell them to the reader, but as I understand the rules of the Demon's Union, no protection is given to a person who does not figure out the answers for himself.

X. Permutations and Combinations

155. FUNDAMENTAL FORMULAS. By a *combination* is meant a sub-group taken out of a larger set of objects, where the identity of the members of the sub-group is important but not the arrangement or order. For example, ABC is a combination of 3 letters out of the 26 in the alphabet; the combination BCA is identical with it.

By a *permutation* we mean a combination in which the *order* as well as identity of the components is important. Thus, BCA is a different permutation from ABC.

The term combination always implies a sub-group out of a larger set; such an expression as "the combination of n out of n objects" means no more than "the set n." But "the permutations of n objects out of n" has real meaning, since the same set of objects can be arranged in different orders.

It is evident that the number of permutations possible in selecting sub-groups is at least equal to the number of combinations, and usually exceeds it. As a rule, permutations are easier to reckon directly than combinations. The fundamental formulas for combinations are derived by reckoning permutations and then dividing by the number of permutations of which each combination is susceptible. In some problems, however, it proves easier to reckon combinations directly and/or to count permutations from this number.

Some of the fundamental formulas are as follows:

$$P^n_n = n! \tag{1}$$

$$P^n_r = \frac{n!}{(n-r)!} \tag{2}$$

$$P^{a,\,b,\,c\,\ldots\,r=n}_n = \frac{n!}{a!\,b!\,c!\ldots r!} \tag{3}$$

$$C^n_r = \frac{n!}{r!(n-r)!} \tag{4}$$

The symbol ! is read "factorial" and means the product of all integers $1\times2\times3\ldots$ up to the specified integer. For example, $5!=1\times2\times3\times4\times5=120$. If you have to expand any of the above formulas by actual multiplication, start by cancelling out the denominator terms (usually the answer must needs be an integer, so that the entire denominator will cancel). For example,

$$\frac{7!}{3!4!} = \frac{7\times6\times5}{3\times2\times1} = 7\times5 = 35$$

Formula (1) gives the number of permutations (P) of n out of n objects. Formula (2) gives the number of permutations of n objects taken r at a time. (3) gives the permutations of n objects (all at a time) of which a are of one kind, b of another kind, c of another kind, etc. Within each kind the objects have no separate identity. (4) gives the number of combinations (C) of n objects taken r at a time.

The counting of permutations and combinations is essential to the solution of many practical problems of diverse types. The best-known type is questions of probability. Before we consider that subject, I will give the reader some exercises in counting. Some of them invoke formulas such as the above; in others, the reader must devise his own formulas. The purely mathematical part of a counting problem is usually easy; the real task is often to see how to *construe* the problem so as to guard against the errors of missing some possibilities and counting others twice over. The subject is proverbially replete with pitfalls. In the computation of card game probabilities, for example, even the most expert mathematicians have made errors of construction, while the odds calculated by an inexperienced person are almost always wrong. A typical error is discussed in the following problem.

156. A COMMON MISTAKE. Smith and Jones had an argument as to the relative probabilities of two types of bridge patterns: 5 4 2 2 and 5 4 3 1. (The four digits in each pattern indicate the number of cards held in a suit. Thus, 5 4 2 2 indicates a holding of 5 cards in one suit, 4 in another, and 2 in each of the other two suits.) They settled the argument by a calculation as follows:

For the 5 4 2 2 pattern we will first reckon all possible combinations of 5 cards out of a suit, which is $\frac{13!}{5!8!}$. The number of combinations of 4 cards out of a second suit is $\frac{13!}{4!9!}$. For the third and fourth suits we have, each, $\frac{13!}{2!\,11!}$. The entire pattern can be formed in

$$\frac{13!}{5!8!}\times\frac{13!}{4!9!}\times\frac{13!}{2!\,11!}\times\frac{13!}{2!\,11!}$$

different ways.

Similarly, for pattern 5 4 3 1 the count of combinations is

$$\frac{13!}{5!\,8!}\times\frac{13!}{4!\,9!}\times\frac{13!}{3!\,10!}\times\frac{13!}{12!}$$

The ratio of the two quantities is

$$\frac{C\ (5\ 4\ 2\ 2)}{C\ (5\ 4\ 3\ 1)}=\frac{3\times 12}{2\times 11}=\frac{36}{22}$$

Smith and Jones agreed that the 5 4 2 2 pattern is more likely to be dealt than the 5 4 3 1 in the ratio of 18:11.

But this conclusion is wrong. The fact is that the 5 4 3 1 pattern is the more frequent. Where did Smith and Jones go wrong?

The basic idea of computing the number of different hands of 13 out of 52 cards that are of 5 4 2 2 and 5 4 3 1 pattern respectively is sound. So is the idea of setting the two quantities in ratio. The error is entirely in the computation.

157. THE ANAGRAM BOX. If a box of anagram letters contains eight G's, nine M's, and thirty each of A, N, R, how many ways are there to pick out and arrange letters to make the word ANAGRAM?

158. MISSISSIPPI. With the same anagram box, in how many ways can you pick out and arrange letters to make MISSISSIPPI? Besides nine M's there are twenty-eight I's, twenty-four S's, and eight P's.

159. POKER DICE. The several varieties of the game "poker dice" are all based on the casting of five (or more) dice, the resultant combination of numbers then being treated as a poker hand.

The prevalent ranking of the various hands, from high to low, is as follows:

Five of a kind
Four of a kind
Full house
Straight
Three of a kind
Two pairs
One pair
No pair

In some circles, the 1-spot denomination is ranked above the 6, instead of below the 2. But even here only two kinds of straights are admitted: high straight 6-5-4-3-2, low straight 5-4-3-2-1.

By "no pair" we of course mean no combination that will place the hand in one of the higher classes.

This ranking of hands is borrowed from poker as played with cards. It is based on the relative chances of being dealt each kind of hand. The total number of different combinations of five cards out of 52 is $52!/5!\ 47! = 2{,}599{,}760$. The chance of receiving a hand of specified type is the ratio of the number of combinations of this type to this total. For example, what is the chance of being dealt a straight flush (including royal flush)? Each suit contains 10 possible straight flushes, which can be topped by any denomination 5,6 . . . . K,A. As there are 4 suits, the total number is $4 \times 10 = 40$. The chance of being dealt a straight flush is then $40/2{,}599{,}760$.

If we were to rank the hands in poker dice by reference to the conditions of dice play instead of card play, we would reckon on the basis of $6^5 = 7{,}776$. This is the total of different permutations that can be cast with five dice of 6 faces each. As an easy exercise, let the reader calculate the number of permutations that fall in each class of hand. Reckon each type separately and check by adding the numbers; the total should be 7,776. (The factorials to be expanded are happily very small. The task is essentially logical: to devise a procedure for setting up each equation with assurance that no combination is overlooked and none counted twice.)

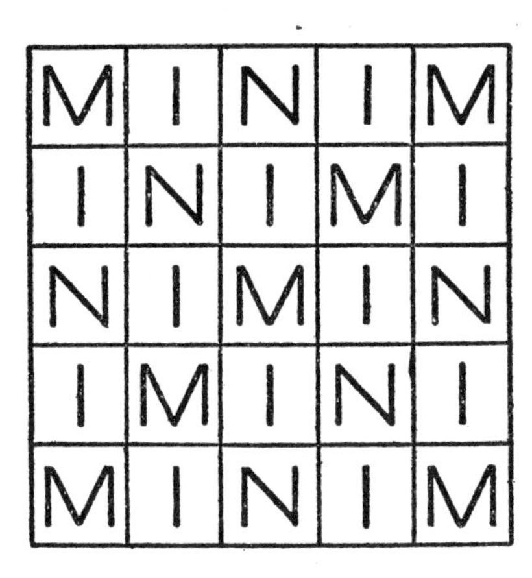

160. THE MINIM PUZZLE. Start on any one of the squares marked M and by consecutive moves to adjacent squares spell out the word MINIM. How many different ways can you do it?

161. THE SPY. The sketch shows the plan of an industrial establishment where some highly secret weapons of war are manufactured. The area is enclosed by a wall containing only three gates, which are guarded night and day.

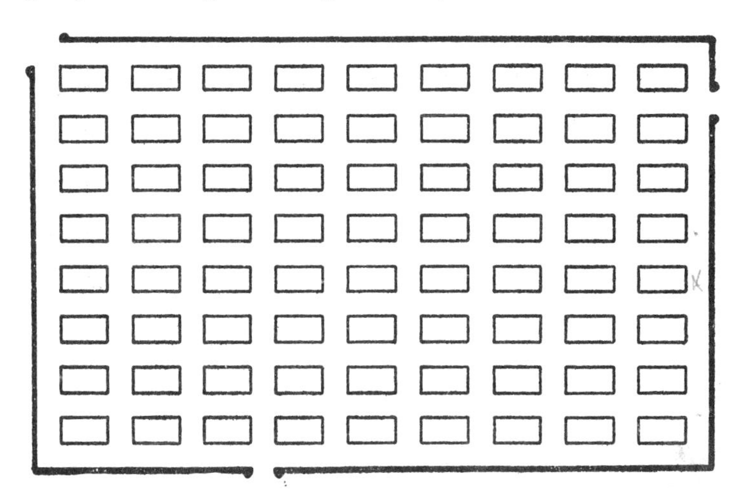

The security officers discovered that information was leaking out of the plant. Little progress was made in the investigation until the following note was discovered in a trash barrel:

"Meet me every Tuesday at the corner near my office. Enter by northwest gate and take a different route each time. That gives you 715 choices."

It was inferred that this note was sent by some spy in the plant to a courier, who was to receive information and convey it outside the plant. In an effort to track the spy, the assumption was made that the courier, on entering by the gate in the upper left of the sketch, would proceed always east or south, never going backwards to west or north. On this assumption, the corner that can be reached in just 715 ways was easily calculated, and a surveillance kept on this corner was indeed successful in spotting the conspirators.

What was the corner where the meetings took place?

162. HOW MANY TRIANGLES? If every vertex of a regular octagon is connected with every other, how many triangles will be formed?

163. THE COIN DROPPER. While the streetcar was stopped at a red light, the operator jingled the silver in his coin dropper. Alarmed by the sound, he dumped all the coins out, counted them, and shook his head doubtfully. It was evident that there had been a run on his change, and he was wondering whether it would last out the trip.

This incident suggested a puzzle to me: In how many different ways could the operator's small change become exhausted?

The coin dropper comprised 4 cylinders. A push on a thumb-lever at the bottom of a cylinder would allow the lowermost coin inside to drop into the operator's hand.

We will say that the first cylinder contains 4 nickels; the second cylinder, 3 nickels; the third cylinder, 5 dimes; and the fourth cylinder, 2 quarters. The question is: In how many different orders can these 14 coins be taken one-by-one from the coin dropper?

Each coin has individual identity. That is to say, taking 2 nickels in succession from the first cylinder gives a different order from taking 1 from the first and then 1 from the second. And of course we must assume that the hapless operator receives no additional change during the process.

164. ROTATION POOL. A professional pocket billiard player practices every day by playing a kind of rotation game. The balls are numbered from 1 to 15, and on the wall hangs a rack with 15 numbered cubicles. As each ball is pocketed, it is removed from the pocket and placed in the cubicle of same number. Just to make it harder, the player stipulates that the row of balls in the rack must never at any time show an interior gap. Thus, if the 6 is pocketed first, the 5 or the 7 must be pocketed next. There is free choice of which ball to sink first, but after that the choice is restricted to balls numbered in sequence with those already in the rack.

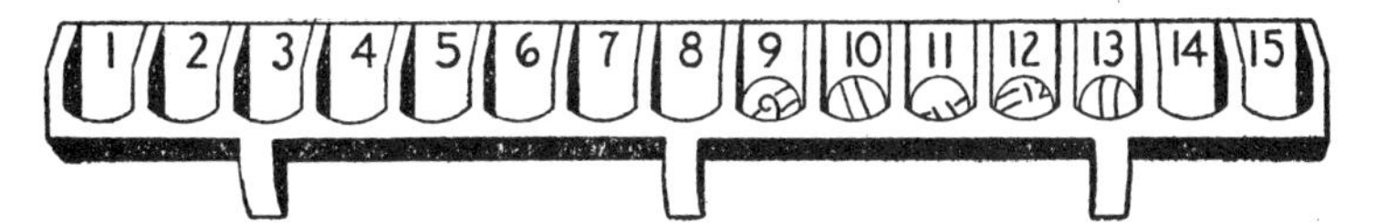

The question is, in how many different orders can the player clean the table of all 15 balls?

This problem was once posed to me by a mathematics teacher, who stated that he knew the formula for the answer but did not see any easy way to calculate it. The trouble is that if you commence counting the number of choices open after each play, you run into different circumstances. For example, if you pocket the 2 first, you then have choice of 1 or 3. If you choose 3, you again have two choices. But if you choose 1, there is no further choice at all. Again, if you start with 7 or 8, you must continue to have choices for some time, but the point at which you cease to have a choice depends on the particular order up to that point.

As in many permutation problems, there is here a way of construing the question so as to make calculation of the answer absurdly simple.

165. THE NECKLACE. Eloise has a quantity of glass beads in four colors, red, yellow, green, and blue. She amuses herself by stringing them on wool yarn in various designs. One of her favorite designs is a necklace of 20 beads, in blocks of 4 of a color.

How many different patterns can Eloise make on this plan? She must use the beads in blocks of 4 of the same color, but we will not insist that she make adjacent blocks of a different color. She may make the entire necklace of one color if she chooses. Or she may use two or three or four colors.

166. TOURNAMENT SCHEDULES. One evening I visited one of the best-known chess clubs in this country with the intention of participating in the weekly "rapid transit tournament." In rapid transit, a player may deliberate no more than 10 seconds before each move.

The entrants being assembled to begin, the tournament director assigned a number to each—but then came a halt. "Where is the book?" he inquired. Search began. "The book! The book!" went hue and cry through the rooms. To the entrants the director announced placidly, "We'll have to call off the tournament if we don't find the book." But presently The Book was unearthed, and the play could commence.

This precious tome, it appeared, was a shabby black notebook into which had been laboriously copied the various schedules for different numbers of entrants. The director was wont to find the proper schedule in this book and call out the pairings of players before each round.

I inquired why the book was deemed so invaluable, and was seriously informed that the chess tournament schedules, devised by some bygone mathematical prodigy, have been published only in an obscure volume obtainable only at the largest public libraries.

I could not help being amused. At a conservative estimate, there are 15,000,000 persons in the United States who can tell you where to find any schedule a chess tournament might need, within a few minutes. Just step to the nearest phone and call up a bridge club.

The widespread popularity of duplicate contract bridge has led to the publication and sale of thousands of schedule cards and books containing schedules. Any one of the round-robin pair or individual schedules will serve for a chess tournament as well.

What is needed in chess is a list of pairings whereby each of n players can be opposed just once to every other. The requirements of a bridge schedule are much more exacting. In a round-robin pair schedule, every pair has to be opposed to every other pair just once, and at the same time a set of duplicate boards must be routed so that every pair may play every board. In an individual schedule, every player must be paired once with every other as partner, must meet every other twice as an opponent (preferably once on his left and once on his right), and a set of boards also must be included in the movement.

The making of such schedules, which involve fundamental problems of combinatorial analysis, has engaged the attention of many eminent mathematicians. Interest in the subject was spurred by the propounding in 1850 of P. T. Kirkman's famous "Problem of the School Girls." In its simplest form, this problem states that a schoolmistress was in the habit of taking her girls for a daily walk. The girls were 15 in number, and were arranged in 5 rows of 3 so that each girl might have 2 companions. The problem is to dispose them so that for 7 consecutive days no girl will walk with any of her school-fellows in any triplet more than once. Several types of solutions were developed for $n=15$, and also for many other multiples of 3. The inquiry was extended to square numbers, and then to multiples of 4. The literature on the subject has grown to voluminous size without exhausting the field, for it is found that there is no general solution for groups of r out of n objects—only particular types of solutions available for n of certain forms.

Edwin C. Howell, an American mathematician, devised pair schedules for whist (and bridge) for any number of pairs from 3 up to 46. The legend devoutly believed by the bridge world is that while working on a schedule for 47 pairs he went mad—and why not!—but I have not found authority for the story.

It is not my intention to pose the reader such a problem as will bring Howell's fate upon him. I am satisfied to point out the extraordinary fascination of the subject of combinatorial analysis. And to encourage the reader to rediscover some of the basic principles, I will ask him to solve the simplest of problems.

We will suppose that the chess club has lost its little black book and has never heard of bridge schedules. Nine entrants are desirous of playing rapid transit. The contest must be scheduled so that during 9 rounds each player meets every other once and has one bye. Also, each player must have White 4 times and Black 4 times. Won't you make such a schedule for the club?

167. PHALANXES. Little Wilbur, a precocious child whom we have met before (in No. 137—*Little Wilbur and the Marbles*), has a number of lead soldiers, which he likes to arrange in rectangular phalanxes. No doubt he is working out the answer to the following puzzle.

With just a dozen soldiers, we can form two different phalanxes 6×2 and 4×3. We may count 2×6 and 3×4 as different, since the width of rank and depth of file are distinct dimensions. We might also count 12×1 and 1×12—all the soldiers in one rank or one file. Altogether we see that there are 6 possible phalanxes with the 12 soldiers.

The number of phalanxes is evidently determined by the number of factors in the total. Suppose that s, the total of soldiers, is composed of $abcd$, four different prime factors. Then counting the phalanxes is merely a matter of counting the combinations:

$$abcd \times 1$$

$$abc \times d$$
$$abd \times c$$
$$acd \times b$$
$$bcd \times a$$

$$ab \times cd$$
$$ac \times bd$$

With only four factors, it is easy to write out all the combinations. But for larger numbers we need a formula that will enable us to compute the total directly. Hence:

How many phalanxes can be made out of s soldiers, if s is composed of n different prime factors?

XI. Problems of Probability

168. PROBABILITY. What is the probability that two cards drawn at random from a full deck of cards will be a pair?

To answer this question, we first count the number of different combinations of two cards that may be drawn:

$$C_2^{52} = \frac{52!}{2!\ 50!} = 1326$$

Then we compute the number of these combinations which are pairs. There are four cards of each denomination; six different pairs can be made out of the four cards. As there are thirteen denominations, the total number of pairs is $6 \times 13 = 78$. The probability that the two random cards will be a pair is given by the ratio $\frac{78}{1326} = \frac{1}{17}$.

To generalize: If s is the total number of events, one of which must occur, and f is the number of events that fall into a class X, then the probability that an event will be of class X is the ratio $\frac{f}{s}$. The set s must be a complete list of *all possible* events, and it must be mutually exclusive—the occurrence of any one event must exclude the occurrence of any other.

Any probability less than certainty is a fraction less than 1. The probability that an event will *not* be of class X is $1 - \frac{f}{s}$.

This follows from the definition of s as a set of events, one of which *must* occur.

Familiar as are these theorems, there is one point that escapes many persons. All mathematical calculation of probability is *deductive:* it deduces the consequences of an initial assumption as to probability. How the assumption originates, by what tests it can be validated—these questions lie beyond mathematics. Computation can no more create the initial hypothesis than geometry can create its postulates.

Statements of probabilities are quite generally made without accompanying statements of their assumptions. But that does not mean that the assumptions are not there. It usually means that the assumptions are evident and are generally accepted, e.g., statement of odds in throwing dice, in the distribution of cards, in drawing colored balls from an urn. But sometimes it means that the statement of chances rests on an implied hypothesis which the speaker himself would reject if it were made explicit.

Failure to scrutinize assumptions raises many a tempest in a teapot. A few years ago a controversy arose over certain probabilities affecting contract bridge. The nugget of the argument is given in No. 169—*That King of Clubs!* Perhaps the reader would like to settle the argument for himself.

169. THAT KING OF CLUBS! Smith and Jones were partners in a game of contract bridge. Smith became declarer at three No Trumps, neither opponent having bid. North made an opening lead of the 3 of Spades, dummy went down, and these were the cards in sight to declarer:

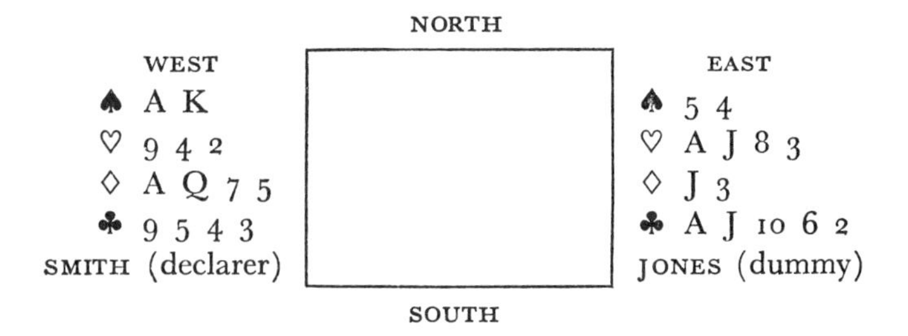

South played the J of Spades, and Smith won with the King. Smith led the 3 of Clubs, North followed with the 7, dummy the 10, and South won with the Queen. Back came a Spade from South, knocking out Smith's Ace. Smith led the 4 of Clubs, North played the 8, and then Smith went into a huddle. Finally he put up the Ace from dummy, but South discarded a Spade instead of producing the K of Clubs as declarer had hoped. Another round cleared the Clubs, but then North cashed three Spade tricks, defeating the contract. Then the argument started.

Jones: Why did you refuse the second Club finesse? You double-crossed yourself.

Smith: I figured that the odds were on South to hold the King of Clubs.

Jones: How do you figure that?

Smith: On the second round, after North played the 8 of Clubs, he had nine cards left, while South still had ten. So the odds were 10 to 9 that South held the King.

Jones: That's a cockeyed argument, and you know it! If that were so, the chances would always be better that fourth hand rather than second hand holds any given card. You know perfectly well that before you touched the Clubs the odds were better on the double-finesse than on swinging the Ace, to lose only one trick.

Smith: Maybe so, but the odds changed after the first Club round.

Jones: How so? All the first round showed, and North's play of the 8 on the second round, was that you would have to guess. If anything else would have happened, there would have been no problem.

Smith: Still, the odds must have changed, because the cards that fell on the first round excluded some combinations that *might* have existed before it was played.

Jones: No, the calculation of the initial chances count only the relevant cases, where you will have to guess. The first Club round simply showed that this was a relevant case.

Smith: Well, I'm not at all sure that the initial chances are better for the double-finesse, anyhow.

Jones: Let's look it up.

Forthwith the players referred to a compilation of card probabilities, and found the following passage:

"With a suit A J 10 x x/x x x x, the defender over the Ace was dealt

1. King singleton in 62/1000 cases;
2. Queen singleton in 62/1000 cases;
3. King-Queen blank in 68/1000 cases.

"A first-round finesse is indicated, since it is as effective as the play of the Ace in these three cases and is superior in the cases where the defender under the Ace was dealt both missing honors plus one or both of the small cards. The double-finesse is seen to be superior to playing the ace by 124:68."

Smith: The book is wrong on its own facts! When I took the first Club finesse and South played the Queen, I knew that Case 1 here—singleton King—did not exist. That left the odds 68 to 62 that South was dealt King-Queen rather than blank Queen. So my play of the Ace was correct!

What says the reader? Is Smith right or wrong? Assuming the statistics on the splits, as given by the book, to be correct, do the odds favor the play of the Ace or the Jack on the second round of the suit, after the first finesse has lost to the Queen?

170. ODDS. The likelihood that an event will occur may be stated either in terms of *probability* or in terms of *odds.* Both methods are widely used. Confusion sometimes arises because the language of a statement does not make clear which point of view is intended. *Chances for* and *chances against* are ambiguous terms; only the context can show whether they mean probability or odds.

If the *probability* that an event will occur is $\frac{f}{s}$, then the *odds against* it are $s-f$ to f, and the *odds for* it are f to $s-f$. (Odds are usually written as ratios and probabilities as fractions.)

The language of odds is favored by bettors, for whom it has an obvious advantage. If a bookmaker "lays" odds of 3 to 2 on Pacemaker to win, a bettor who puts $2 on Pacemaker will receive $3 if the horse does win. But in bookmaking the odds have to be translated into probabilities (however much the fact is disguised by algebraic short cuts). For example:

In a three-horse race, a bookmaker lays odds of 2 to 1 on Agamemnon and odds of 3 to 2 on Behemoth. What are the correct odds on Calypso, if the bookmaker does not give himself a percentage?

171. PARLIAMENT SOLITAIRE. Here is an easy question concerning the patience game Parliament (also called Tournament).

Two 52-card decks are shuffled together and 8 cards are dealt face up. If not a single ace or king appears in this layout, the game very probably cannot be won, so that it is quite proper to pick up the cards, reshuffle, and try again. That is all you need to know about Parliament to answer the question:

What are the odds that you will turn up at least one ace or king in the first 8 cards?

As the factorials involved are rather large, you need not expand them. Just show the formula and make an estimate of the result.

172. EVERY THROW A STRAIGHT. A die has six faces, numbered from 1 to 6. Your chance of throwing any given number, say 4, is $\frac{1}{6}$—assuming the die to be honest. If you cast two dice together, you have $\frac{1}{6}$ chance of getting a 4 on the first, plus $\frac{1}{6}$ chance of a 4 on the other, or $\frac{1}{3}$ chance of getting at least one 4. By increasing the number of dice you roll simultaneously, you similarly increase your chances of rolling at least one 4. In fact, if you roll six dice, your chances are $6 \times \frac{1}{6}$ or 1, which is certainty. But what goes for 4 must go for any other number on the dice. It is likewise a certainty that you will cast at least one 6, one 5, and so on. In fact, any roll of six dice together must result in a straight . . . help! What's wrong here?

173. TREIZE. In the French gambling game Treize, a deck of 52 cards is shuffled and the cards are turned up one at a time. As they are turned, the dealer counts "One, two . . ." up to thirteen, then again from one to thirteen, and so on four times. If the denomination of a card coincides with the number called, that fact is a "hit." The gamble enters by way of bets on how soon a hit will be made, or on the failure to make any hit. The ace is counted as one; jack, queen and king respectively are eleven, twelve and thirteen.

I have seen the game played as a solitaire. The player counts "Ace, two, three . . . jack, queen, king" four times as he turns the cards—provided that he is lucky enough to get that far; the solitaire is deemed to be won if the player gets through the entire deck without a hit.

It would be easy to ask: What are the odds on winning Treize solitaire? but I shall not do so. The huge numbers involved are not enticing. But suppose we simplify the problem. Let us take only 6 cards, numbered from 1 to 6, shuffle them, and play Treize. What are the odds we will get through the half-dozen without a hit?

As a matter of fact, the odds when the full deck is used are not much different. The change in odds as the number of cards is increased over 6 is extremely slow.

I warn the reader that the problem is rather more involved than may appear at first sight. It is easy to figure the total permutations of 6 cards—factorial 6. But when it comes to counting the permutations in which no card occupies the ordinal position of its own denomination, no simple formula can be evolved. In fact, the task is like that of counting partitions: the practical method is to construct a table for the required permutations out of $n!$ as n takes the values 2, 3, 4, 5, 6. Each entry in the table is based on the previous entries. In making the table the solver should discover the principles whereby it can be extended infinitely; my object in posing the problem is to give him this exercise in induction.

XII. Number Games

174. MATHEMATICAL GAMES. The very notion of a "mathematical game" is a contradiction in terms. The moment every factor in a contest yields to precise mathematical calculation, it ceases to be a game. Yet history affords numerous examples of completely exhaustible "games" of undiminished vitality: the schoolboys who played tit-tat-toe on the steps of the Acropolis have their counterpart today in every land and every walk of life.

Good reason can be adduced why such games never die. They are usually played as a challenge to the uninitiated to discover "the secret" of winning (or avoiding loss). Thus they encourage reflection and analysis, and are more apt to give instant reward than many other activities of mind, because their "laws" are not far to seek.

It might be expected that study of those games whose merits are wholly known might shed light upon others still in some degree incalculable. Such is indeed the case. Games of a purely intellectual character, e.g., checkers, are gradually being taken from the domain of art into the domain of science. Some persons deplore this evolution, but why worry? The human mind can devise new games much faster than old ones are exhausted.

The games I shall discuss are of two types, numerical and tactical, dealing respectively with cardinal numbers and with relations of position. All are played between just two contestants. The focus of our inquiry is whether the first player (he who moves first) or the second player must win with the best of play on both sides, and what this best play is.

175. ONE PILE. This purely numerical game has been traced back to remote antiquity, and probably it antedates the games of position, such as tit-tat-toe.

A number of pebbles or counters of any description is massed in one pile. The two players draw alternately from the pile, the object being to gain the last counter.

If it were permitted to seize the whole pile, the first player would of course win; if the draw were limited to one counter at a turn, the result would depend upon whether the number in the pile were originally odd or even. Therefore, a minimum draw of one counter is set, with a maximum greater than one.

Suppose the limits are 1 to 3 counters. Then if a player finds just 4 counters left in the pile, he loses. Whatever he takes, his opponent can take the remainder. It is readily seen that the number 4 is a critical one because it is the sum of the minimum and maximum limits of the draw.

In order to leave his opponent with 4 counters to draw from, a player must previously have left him 8. Whether he then drew 1, 2, or 3, it was possible to reduce the pile to 4. Evidently the series of winning combinations, each of which is a number to be left in the pile for the opponent to draw from, is simply the multiples of 4.

If we denote "a winning combination" by w, and the least and most that may be drawn at a turn by a and m respectively, then

$$w=(a+m)n$$

where n is any integer. This formula is quite general, and is independent of the number of counters originally in the pile. If this number is of form w, the first player loses; if it is not, he wins by reducing it to w.

176. TO LEAVE THE LAST. The game can also be played with the object of forcing one's opponent to take the last counter. I leave it to the reader to write the formula for w in this case.

177. TO WIN THE ODD. A more complex form of the one-pile game puts an odd number of counters in the pile, fixes limits on the draw, and gives victory to the player who owns the odd number of counters after the common pile is exhausted. What is the formula for *w*?

178. THREE-FIVE-SEVEN. Kindred games can be played with more than one common pile of counters. A very widespread variety is "3-5-7." Three piles are set up consisting of counters in these amounts. At each turn a player may draw any number of counters from one pile—the whole pile, if he wishes—but may not draw from more than one pile at a time. The usual stipulation is that he who must draw the last counter loses.

Which wins, first or second player, and how?

179. THE THIRTY-ONE GAME. In the smoking room of the liner, Bill Green fell into conversation with a man who introduced himself as Jack Smith. The talk presently turned to mathematical games, and Smith said he knew a puzzle that was rather interesting. From a deck of cards he removed the A, 2, 3, 4, 5, 6 of each suit, and laid the 24 cards face up on the table.

"Now the idea," he said, "is that you turn over a card and then I turn over a card, and so on. We add up the cards we turn as we go along, and we can't go beyond 31. Whoever turns the last card to make exactly 31 wins. The ace, by the way, counts as one."

Bill Green realized that there must be some mathematical principle in which cards to turn, but he had no objection to trying the game in fun. Several games were played, some won by Smith and some by Green. Smith didn't seem to play by any particular system, but Bill noticed that whenever he made the total 24 he won. It dawned on Bill that 24 was a magic number, for, having to add to it, a player could neither reach 31 nor prevent his opponent from reaching it at his next turn. By the same reasoning 17 was a magic number—7 less than 24. In fact, the whole series of numbers made by subtracting 7's from 31 down were winners.

Bill tried out his theory by beginning with a 3. Smith turned a 5 and Bill answered with 2, to make 10. He won the game by sticking to the series 3, 10, 17, 24.

Smith cogitated over this result and remarked, "I think I see it now. I bet I win the next game."

Forthwith he began with a 3, and Bill at random turned a 6. Smith triumphantly chose 2, and didn't seem discomfited when Bill turned another 6. But he was indeed crestfallen when Bill won again.

"I guess I made a mistake," he said. "I thought I had the hang of it. After you took the first 6, let me see—oh of course, I should have taken a 5!"

Bill was inclined to dispute this and to explain the simple formula, but Smith cut him off.

"Don't tell me. I'm sure I get it now. I'll bet you that I win the next game!"

Bill Green was too wary a bird to bet with a stranger. Still, it was his turn to play first, and Smith evidently hadn't grasped the real idea, and—well, somehow Bill agreed to a bet that was a little more than he could afford to risk.

Bill began with a 3, and Smith turned a 4. To stay "in the series" Bill turned another 3 to make 10, and Smith turned another 4 to make 14. The play continued in the same way, Smith turning only 4's and Bill turning 3's to stay in the series. When Smith turned the last 4, the total was 28. The 3's being all gone, Bill had to turn an A or 2, and his opponent won.

Bill Green is not the first "gull" who has fallen into a swindle that is hoary with age. The game of Thirty-one has been used to fleece many persons who have the proverbially dangerous "little knowledge."

The game is indeed analogous to No. 175—*One Pile,* and it is true in general that

$$w=31-(a+m)n=31-7n$$

But here exists an added feature, a limitation on the number of times the same integer may be chosen. It is not feasible to seize the *w* series at once.

The first player can nevertheless force a win. How?

180. THIRTY-ONE WITH DICE. A clever variant of the Thirty-one Game, invented by a conjurer, has proved baffling to many "sharks" who thought they understood all about mathematical games.

This variant is played with a single die. The starting number is fixed by a chance roll of the die. Thereafter each player gives the die a quarter-turn, in any direction he pleases, to bring a new number uppermost. A running total is kept of the numbers so turned up, and he wins who reaches the total 31 or forces his opponent to go over 31.

The die variant is actually very different from the game with 24 cards. There, only four duplicates of each digit exist, but all unturned digits are available to the player in his turn. Here, the number of digits is unlimited, but at every turn the player finds two of them unavailable—the number already up, and that on the opposite face.

The numbers on opposite faces total 7, so that the player has a choice of only two pairs of numbers out of the couples 6—1, 5—2, 4—3.

What number or numbers, turned up by the random roll, spell victory for the first player? What is the system whereby he can preserve his advantage and force the win?

XIII. Board Games

181. THE PRINCIPLE OF SYMMETRY. The number games discussed in the previous chapter point to a general law which I shall call *the principle of symmetry.*

The applications of the principle to some games, e.g., chess and checkers, are well known, but nowhere have I seen the principle enunciated as such. Yet it pervades most games of a mathematical nature, both numerical and tactical.

This is the principle:

The player who can present his opponent with a symmetrical configuration thereby gains an advantage. His opponent has to disturb the symmetry and thereby leave the first player with the option (a) of restoring the symmetry by a cognate move, or (b) increasing the asymmetry by another move.

In this definition, *symmetry* must be construed broadly to mean a perfect correspondence of pairs of entities, whether they be magnitudes (number games) or points (tactical games).

Let us review briefly some of the best-known applications of the principle of symmetry.

Three-Five-Seven

We have seen that any array is a *w* (winner for the player who presents it to his opponent) if every component power of 2 is represented an *even* number of times. Such an array is symmetrical in the sense defined above. Adherence to the principle is the one and only requisite; the player who has achieved a *w* continues to restore the symmetry at every turn until his last, when an unsymmetrical play clinches the victory.

Checkers

The principle of symmetry operates in what is called "the move." The basic proposition of the move is shown in Fig. 1. The configuration is symmetrical; the player whose turn it is to move is, according to principle, at a disadvantage. For Black, this disadvantage would be fatal: compelled to retreat, he will be driven to the side of the board and captured. For White, the disadvantage would be that he cannot win: he is saved from loss by a peculiarity of the arena, the double corner.

The rule-of-thumb for calculating the move shows clearly the domination of the principle of symmetry: Count all the pieces in one *system* or the other (when Black and White each have the same number of pieces). If the total is even, the player whose turn it is to move has "the move" against him. If the total is odd, the player whose turn it is to move has "the move."

Fig. 1
Black to move loses
White to move draws

(By a *system* is meant the 16 playing squares of the board that lie on four alternate files, e.g., the squares 5—13—21—29, 6—14—22—30, 7—15—23—31, 8—16—24—32. The two systems are conventionally designated as Black and White according to the side of the board on which the squares abut.)

"The move" is not the whole of checkers, although it is controlling in much of end-play. Other factors also operate, generated by (a) the character of the arena (asymmetry of adjacent corners, one being double and one being single), and (b) dynamics of the pieces (an ordinary *cut* changes the move but *not all* configurations of capture do so).

It may be said that the vitality of any mathematical game depends upon suchlike peculiarities of arena and dynamics whereby the tyranny of the principle of symmetry is to some degree circumvented!

Chess

The principle of symmetry operates in what is called "the opposition." The basic position is shown in Fig. 2. The rule for

Fig. 2

Black to move loses
White to play, Black draws

the opposition of kings is that the kings are opposed when they stand on squares of the same color and an odd number of ranks or files intervene between them. In the "near opposition" shown in the diagram, the player whose turn it is to move finds the opposition against him: his king must give way. The enemy king then has choice of resuming the opposition or of making the "passing move."

Thus, with Black to move: 1—Kd7; 2—Kf6. The White king passes and threatens to reach f7, whereupon Black will no longer be able to block the advance of the pawn. Black has to play 2—Ke8, but then with 3—Ke6 White again seizes the opposition. Again Black has to give way, and White will then be able to pass to f7 or d7.

If White is to move, he cannot by king moves alone compel Black to give way. Thus 1—Kf5, Kf7. Black simply maintains the opposition. To change the move, White tries 2—Pe5. But now, owing to a peculiarity of dynamics, the opposition is insufficient to win for White. The Black king is indeed forced back: 2—Ke7; 3—Pe6, Ke8! (Kf8 loses); 4—Kf6, Kf8; 5—Pe7 check, Ke8; 6—Ke6. Black has no legal move, is "stalemated." By the rules of the game a stalemate is a draw.

As in checkers, the opposition usually becomes of importance only in end-play. In chess the principle of symmetry enjoys much less dominance, owing to the much greater dynamic complexity of the pieces.

182. SAM LOYD'S DAISY PUZZLE. An elegant application of the principle of symmetry was propounded by Sam Loyd in his *Daisy Puzzle*. As he tells the story, he was one day touring in the Swiss Alps and came upon "a little peasant girl gathering daisies. Thinking to amuse the child, I showed her how to prognosticate her matrimonial future, by plucking off the petals of the flower. She said the sport was well known to the country lassies, with the slight difference that a player was always at liberty to pluck a single petal or any two contiguous ones, so that the game would continue by singles or doubles until the victorious one took the last leaf and left the 'stump' called the 'old maid' with your opponent.

"To our intense astonishment the pretty mädchen, who could not have been more than ten years old, vanquished our entire party by winning every game, no matter who played first."

Naturally, the maid must have depended, in one case or the other, upon her opponent's ignorance of the game! The question is, which player can force a win, the first or the second? I shall not spoil the reader's enjoyment of this beautiful puzzle by giving the answer here.

The daisy illustrated by Loyd, it must be stated, has thirteen petals.

183. DUDENEY'S CIGAR PUZZLE. It would be a wonderful thing to have a book of puzzles, all of which could be solved by "common sense," requiring no knowledge of formal mathematics.

But that book probably never will be written.

The trouble is that, when you foreswear all knowledge that comes through schooling, there is precious little left in the pristine mind on which to base a puzzle.

Through the ages, a small store of "common sense" puzzles has accumulated. The prize of the collection, in my opinion, is Dudeney's Cigar Puzzle. I cannot resist narrating that I first became acquainted with this gem while reading in bed, and that like an illustrious precursor I startled the household by jumping out of bed, dancing about and crying "Eureka! Eureka!"

Here is the puzzle:

"Two men are seated at a square-topped table. One places an ordinary cigar (flat at one end, pointed at the other) on the table, then the other does the same, and so on alternately, a condition being that no cigar shall touch another. Which player should succeed in placing the last cigar, assuming that each will play in the best possible manner?"

The supply of cigars is supposed to be inexhaustible, and the cigars are supposed to be uniform in size and shape.

184. THE CARPATHIAN SPIDER. The Carpathian spider has odd habits. Having built a web, it retires like any normal spider to a place of concealment. But when a fly becomes enmeshed, the Carpathian spider does not at once run to deal its lethal sting to its prey. It advances only a little distance, then

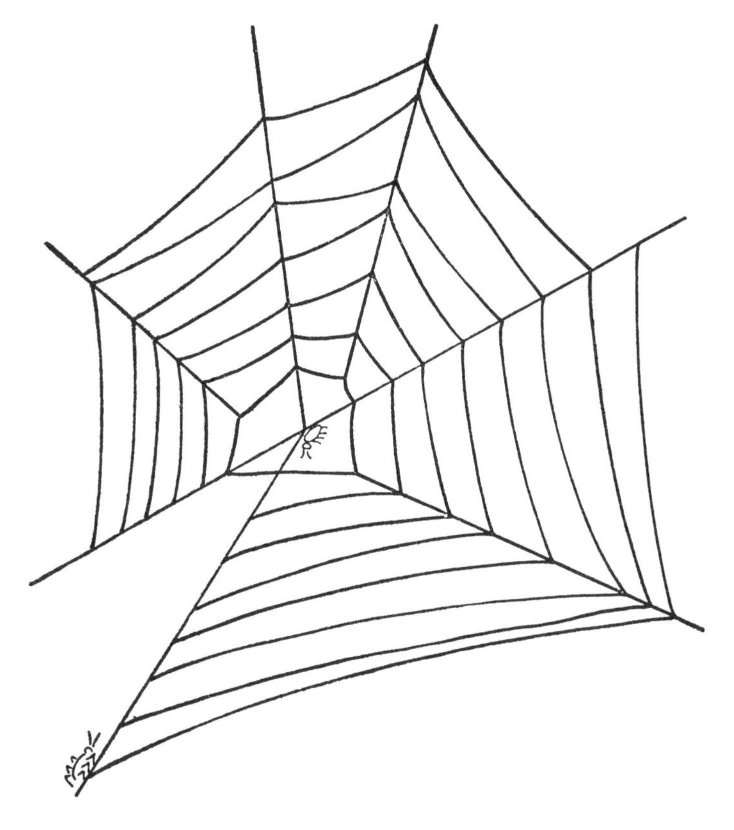

stops and scrutinizes its prey to estimate whether there is danger that it will escape. Then it advances again and stops again, thus proceeding by stages to the kill. Perhaps a part of its motivation is a sadistic delight in applying the torture of suspense to its victim.

The following account of a battle between a Carpathian spider and a fly was narrated by a witness in whom I have the utmost confidence.

The scene at the outset of this remarkable fray is shown in the illustration. The spider was lurking at an outermost point of his web when a fly lit upon the central point. Immediately the spider advanced along the radial strand, going only as far as its intersection with the next transverse strand. When it stopped, the fly, no doubt spurred with new hope, succeeded in crawling away from the center as far as the next intersection on a radial strand. The spider thereupon made a second advance, while the fly, paralyzed with fear, stayed motionless.

And thus the chase proceeded. Each time the spider moved the fly froze, and each time the spider stopped the fly crawled further away. The fly evidently could not release himself wholly from the web, but could manage to bog along the strands. The strange part of the story is that each separate move by either insect took it along a strand of the web from one intersection to the next—never more.

It seemed to the witness that the spider would never catch the fly by such dilatory tactics. But he was wrong. Eventually the spider cornered the fly and was able to move to the same point as its victim, which it then despatched without mercy.

How did the spider win this strange gavotte? Remember that in the situation shown by the illustration the spider made the first move.

185. TIT-TAT-TOE. Probably the most widely-known "game" is Tit-Tat-Toe, or Noughts and Crosses. It is found in every civilized country on the globe, and is known to have been played in ancient times. Probably 99 out of 100 schoolboys investigate its meagre permutations and discover for themselves that with best play on both sides it is a forced draw.

Ingenuous as is this pastime, a study of just why certain moves lose sheds light upon the strategy of certain related games, games not yet completely analyzed.

In Fig. 1 are shown the three possible openings by the first player, Cross. The Noughts in each case show the only replies that draw. The "side opening" is evidently weakest, since there is option of two replies.

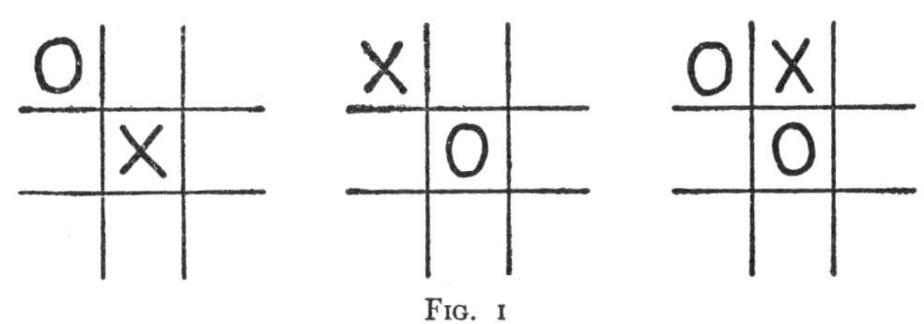

FIG. 1

Any other reply by Nought to the first move loses because it permits Cross to establish a *fork*—the threat of making 3-in-a-row on either of two lines. Nought cannot block both. Now let us make an abstract of the conditions under which a fork can be forced.

Abandoning the mechanics of Tit-Tat-Toe, I will now indicate the points on which marks can be written (or counters placed) by circles, connected by lines to show available winning configurations. Fig. 2 shows what I will call the *critical triangle,* one of the simplest configurations of points and lines on which

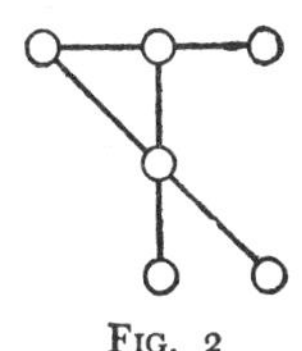

FIG. 2

a fork can be based. The triangle is defined by three lines, each containing three available points of play (where 3-in-a-row wins), with each vertex of the triangle lying on two of the lines.

If Cross, the first player, occupies a vertex of a critical triangle, and if Nought then plays on some other part of the board, Cross wins by playing on a second vertex. Nought has to block the threat on one line, whereupon Cross takes the third vertex and forks across the other two lines.

Therefore, when Cross takes the first vertex, Nought must immediately play on the triangle to avoid loss. Whether he must take a vertex depends upon the limitations of the system: by hypothesis there is more to the board, else Nought could not go wrong. But, it is important to note, the strongest reply of Nought is to *take a vertex*. Here he cuts two lines; elsewhere he cuts but one.

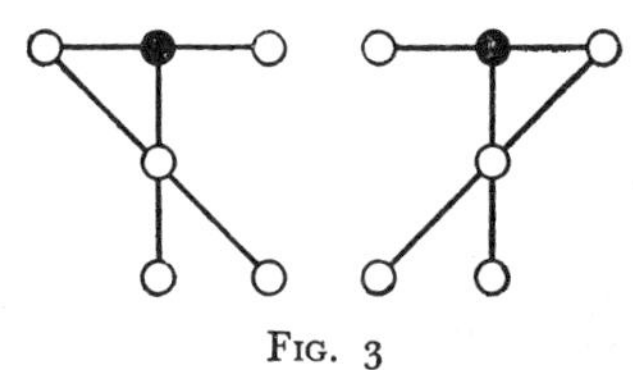

FIG. 3

The "side opening" in Tit-Tat-Toe takes a vertex of two critical triangles, depicted in Fig. 3, where the black dot marks the point seized. If the two triangles were wholly distinct, except for the one common point, then Cross could win by force, as Nought could not play into both at once. As it is, the two triangles here have no less than four points in common, so that seizure of any one of the remaining three forestalls a fork. In other words, a sufficient reply to the "side opening" is to play into either adjacent corner or into center.

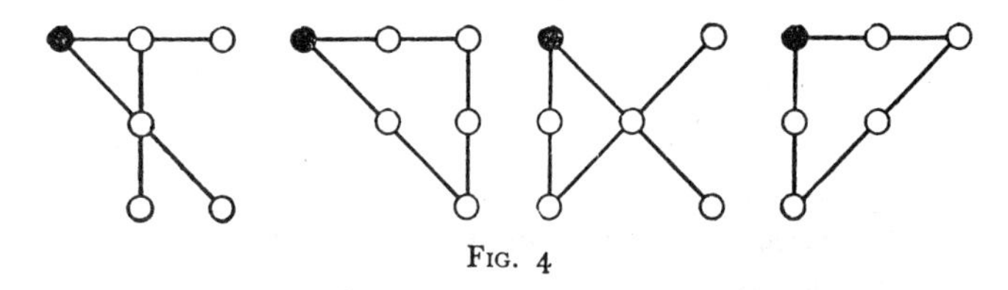

FIG. 4

The "corner opening" is much stronger; this point is a vertex of seven critical triangles. Four of them are shown in Fig. 4; the others are symmetrical reflections of the first three. There are more threats than could be scotched in one turn, were it not for the fact that *the center point of the layout is common to all of them.* Seizure of the center suffices to draw. While this point is not a vertex in *all* the critical triangles, any effort by Cross to utilize triangles where it is not is defeated by the limitations of space. Going back to Fig. 2, we see that Nought can draw by taking any point in the triangle, even non-vertex; for there are only three available lines in all, and Nought can kill each in turn.

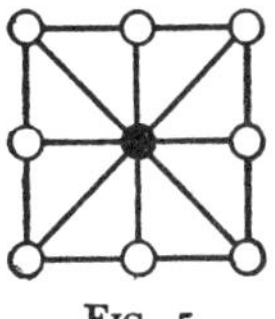

FIG. 5

The "center opening" is clearly strongest. The center point lies on every possible critical triangle that can be picked out of the nine points. To reply by seizing a side is insufficient. In Fig. 5 suppose Nought takes point 2. Then Cross can take 3, forcing his opponent to block at 7, and then a play on either 6 or 9 makes a fork.

A side point lies on only two lines of the layout; a corner point, lying on three lines, must be superior. Seizure of a corner in fact gives Nought a draw. In Fig. 5 let Nought's reply be to take point 1. If Cross plays 2 or 4 he exerts only a "one-move threat," since one of the two lines radiating therefrom is already killed by 1. If he plays upon 3, 6, 7, or 8 he indeed exerts a "two-move threat" of fork, but Nought threatens first. His blocking answer lines up with 1 and compels Cross to block instead of completing the fork, which Nought thus gains time to kill.

186. AN UNLIMITED BOARD. From the foregoing considerations in Tit-Tat-Toe we can deduce that, if the playing field is sufficiently extensive, the first player can always force 3-in-a-row.

Any centrally-located point on a large board has all the advantages of the center point in Tit-Tat-Toe. But more—each line passing through it is in effect two lines, since from the point there is space available on both *rays* to make 3-in-a-row. It is necessarily possible to describe two critical triangles having only one common point, and Nought (the second player) cannot block in both triangles at once. Also, Cross (the first player) can develop two triangles that overlap to a considerable extent, without letting Nought develop such a counter-threat as saves him in Tit-Tat-Toe.

It may well be suspected that, on a board of unlimited extent, Cross can establish by force *a row of more than 3*. Such is actually the case. At least 5-in-a-row can be forced.

Here we have to reckon with a configuration even simpler than the critical triangle—the critical line.

Suppose the game is to make 5-in-a-row. Suppose that Cross gets 4-in-a-row, and that the points at each end of this line are open (unoccupied). Then he must win, since Nought cannot block both ends of the line at once. Thus, what we will call an "open 4" is a critical line.

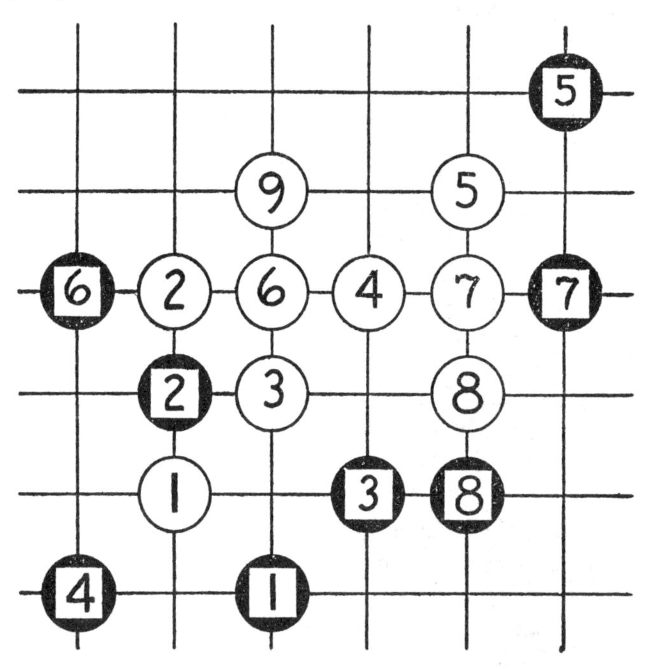

On an unlimited board, the first player can always create a fork with two (sufficiently) "open 3s" and thereby make a winning "open 4." One of many possible variations is shown in the diagram. The moves of each player are numbered from 1 up, the first player being "white."

If the play is followed move by move, it is seen that the second player, in his first three turns, tries to cut as many lines as possible among those radiating from the points seized by the first. At his fourth turn and thereafter he is kept busy blocking open 3s; his only option is which end to block. Despite his efforts to combine defense with counter-threat, the first player has established a fork of two 3s by 9—6—3 and 9—4—8.

The present state of knowledge leaves open the question: Can the first player force 6-in-a-row? For purposes of making an interesting (and playable) game, however, better than extending the winning row is to hedge the board or the rules with certain limitations, as is done in the next two games described.

187. GO MAKU. The oriental GO board is a grid of 19×19 lines. Stones are played upon the intersections, not in the squares. Hence the field comprises 381 points. Two different games are played on this board. One is the game GO, a profound tactical game at least as complex as chess but with practically no literature. The other is Go Maku, also called Go Bang, the object of which is simply to establish 5-in-a-row by the alternate laying down of stones.

Since, as is shown in the preceding section, the first player can always force a win, Go Maku has to employ an arbitrary rule: A player may not so place a stone as to establish a fork of two "open 3s." He may, however, fork an open 3 with a one-ended 4, or make a doubly-open 4.

Through the operation of the rule, a player often finds himself in the exasperating situation of being unable to play on a key square, and of losing the fruit of prolonged efforts. Whether the first player can force a win is unknown—hence Go Maku is still a real game! It is my belief that the game is a draw with best play on both sides. The second player can certainly put up a prolonged defense if he is willing to forego any attempt to win.

At all events, Go Maku offers a fascinating challenge to player and theorist alike.

188. THE MILL. The Mill, also called Nine Men's Morris, is an ancient game that may have had a common origin with Tit-Tat-Toe. The board, shown in the illustration, comprises 24 points, connected by 3 concentric squares and 4 transversals.* Each player is provided with 9 counters, of distinctive color. Each in turn lays a stone on one of the points (line intersections) until all 18 stones have been played. Then each in turn moves one of his stones along any line on which it stands, to the next adjacent point.

The object of play is to decimate the adverse army. Each time a player establishes 3 stones on any line of the board, called a *mill,* he is entitled to remove an adverse stone from the board, with the proviso that he may not take one from an adverse mill. Once a mill is established, the owner may "open" it by moving one stone off the common line, then "close" it by moving the stone back. Even though composed of the same 3 stones, this formation now counts as a new mill and entitles the owner to remove an enemy piece.

In some countries, when a player is reduced to 3 stones they become "wild" and may be moved from point to point regardless of connection by line. This flourish is little more than a solace for the vanquished, for a superior force will usually win against

*Another version of the board connects each triplet of corners by a diagonal line. Whether these lines should be added has long been a matter of controversy. Angelo Lewis (Professor Hoffman), writing in 1894, said that the pattern with the diagonals "is preferred by some players, though the addition is stoutly resisted by the champions of the original game." On the other hand, H. E. Dudeney says (1917) "Sometimes the diagonal lines are omitted, but this evidently was not intended to affect the play: it simply meant that the angles alone were thought sufficient to indicate the points." The present writer holds to the no-diagonal school; for one reason, this is the style in the Scandinavian countries, where the mill game is as commonly played as checkers.

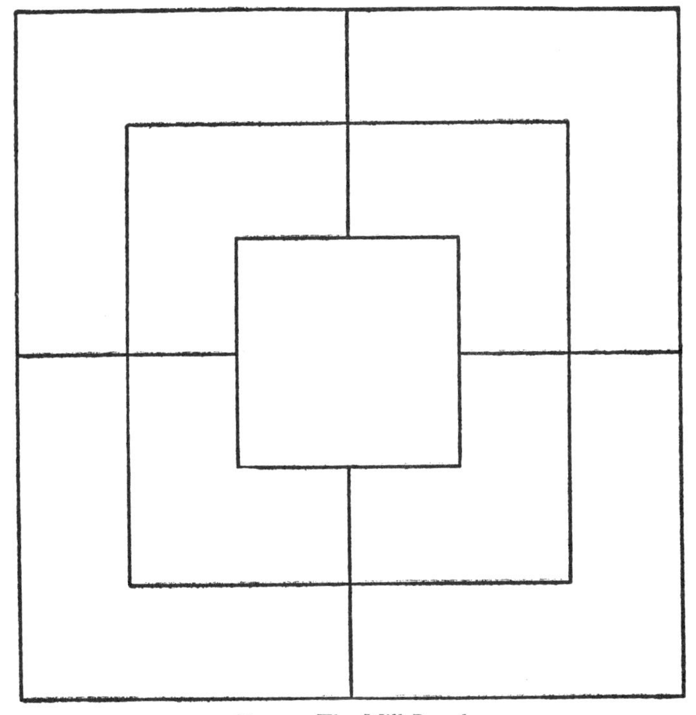

FIG. 1—The Mill Board

the 3. Of course when a player is reduced to 2 stones the game is over.

As a social two-handed pastime, Mill is one of the best of games. In opportunity for skill it stands between Tit-Tat-Toe and Go Maku. Having fewer possibilities than the latter, it makes much less demand on the player's powers of analysis and visualization. Still, it has not been exhausted.

As a subject of mathematical inquiry, Mill is a fascinating combination of ideas. The basic 3-in-a-row principle, completely exhausted in Tit-Tat-Toe, is given vitality in two ways. (a) The board is somewhat enlarged, but is still so confined that the player continually feels "With a little thought I could analyze this game completely!" (b) A game of placement is combined with a game of movement. It is no longer sufficient to be the first to make 3-in-a-row.

In Fig. 2 is shown a game that has been played many times. It illustrates how dangerous is a little learning. White, the first player, correctly occupies a fourway intersection. Black seizes another fourway point. White has studied the properties of the board just enough to know that Black's failure to play next to the first White stone lets White establish a fork. He proceeds to do so. His plays from 2 to 5 keep Black busy intercepting, and the 6th White stone makes a fork. Black blocks at 6 and White

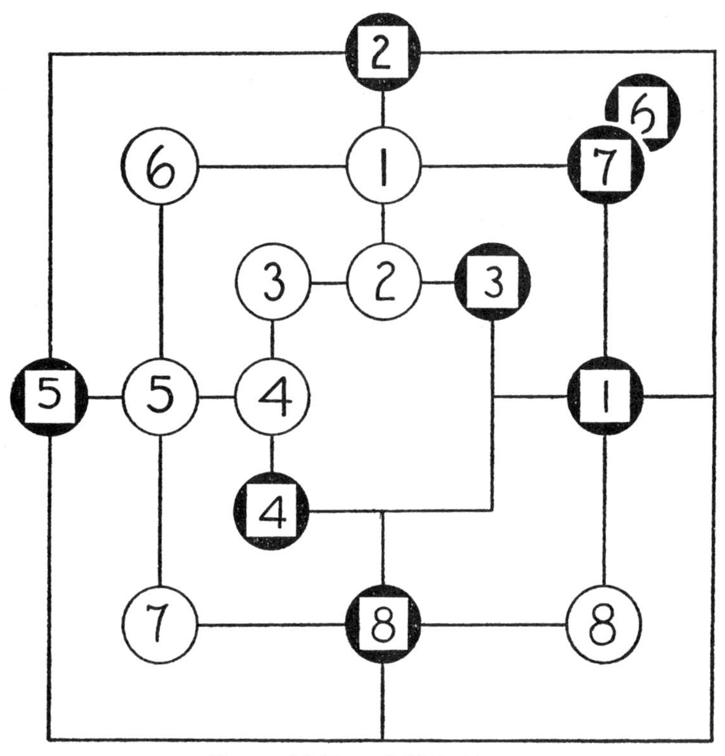

FIG. 2—Making a Mill by Force

makes his mill. Since Black's 6 and 1 threaten to make a mill, White removes one of these stones, say 6. Black's 7th replaces this loss, and compels White to block at 8. Black then has to cut at 8.

Look at the position now. Not one of the White stones can move. Black's 7 stones hold all of White's 8. The 9th stone on each side remains to be placed. Wherever White elects, Black will place his 9th adjacent. White will have one move at most, whereupon Black can trap this last piece. White loses, because in his turn to play he has no move.

It does not follow from this instance that to force a mill in laying down, when opportunity arises, is necessarily fatal. On the contrary, the player who makes the first mill while still maintaining freedom of movement is likely to win.

I have proved to my satisfaction that the game is a forced draw if neither player takes risks to win. But there is remarkable scope for ingenuity in setting traps.

189. SALVO. An excellent two-handed game that combines luck and skill in sociable proportions is Salvo.

Each player uses a pencil and a sheet of "quadrille" paper (lined both vertically and horizontally). Each commences by outlining two squares 10×10 as shown in the diagram. The columns and rows are designated by different indices. The left square is the player's "own" arena, while the other is his opponent's. In his own arena the player places, wherever he chooses, four "ships." Each ship is a series of squares blacked out, in a line vertically, laterally, or diagonally. The "fleet" consists of one "battleship" (5 squares), one "cruiser" (3 squares), and two "destroyers" (2 squares each).

Both fleets being established, the players in turn fire "salvos" into the enemy fleet in an effort to sink it. At the outset, each salvo comprises 7 shots. The player marks 7 of the squares on his enemy arena at which he chooses to shoot, and names them to his opponent, e.g., in the diagram the first salvo was on squares D2, G3, E4, H5, etc. The opponent marks these shots in his own arena. All shots are marked by the number of the salvo, the salvos being numbered consecutively from 1 up so that the player can keep track of the time at which hits were scored.

On completion of a salvo, the player fired at announces whether any of the shots have scored a hit on his fleet. He names the type of vessel hit, but not the square. His opponent of course makes a memorandum of any hits so announced.

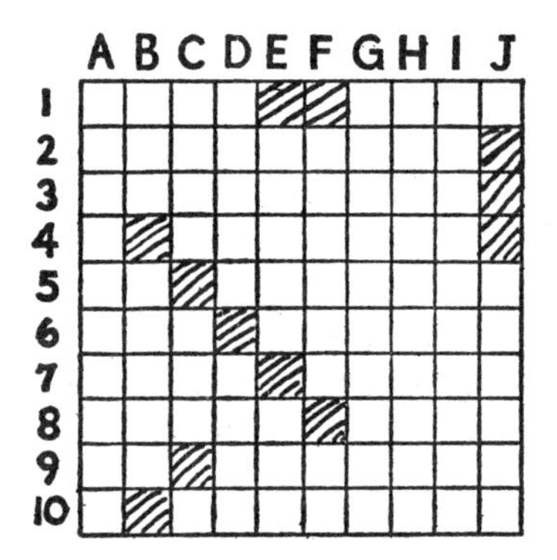

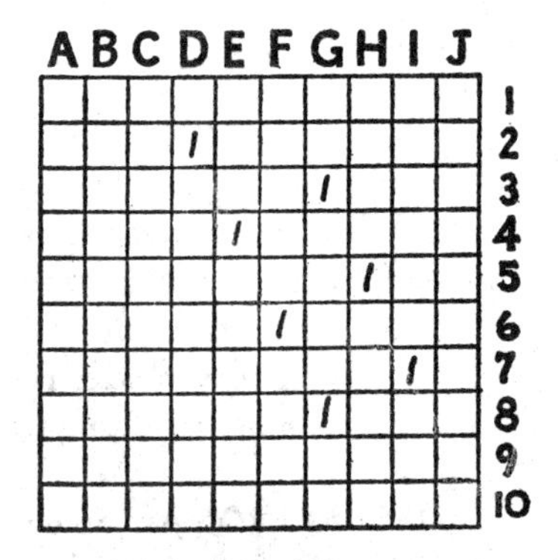

The object of play is to sink the enemy fleet entire by hitting every component square. When any vessel is sunk, the owner's quota of shots in all his subsequent salvos is reduced by 3 for the battleship, by 2 for the cruiser, or by 1 for a destroyer.

There is a premium upon locating the enemy battleship and concentrating salvos to sink it in a hurry. One popular method of commencing is to quarter the enemy board and put the first four salvos in the quarters. A natural formation of shots for the early salvos is shown in the diagram, a kind of chess knight tour. The idea of it is to put one shot in each of several adjacent lines covering groups of lines in all three directions.

The question suggests itself: If your opponent is known to use this formation on his first salvo, where is the best place to put your battleship so as to escape a first-round hit? Also, where is the worst place? Intuition may supply the answers, but give a rigorous mathematical demonstration.

Part Two

SOLUTIONS

Solutions

(Here are the Solutions to the Puzzles in Part One of this book. You will note that the numbers of the Solutions correspond to the numbers of the Puzzles.)

1. HOW HIGH IS A POLE? The pole is 28 feet high. The problem is solved by simple proportion. The height of the pole is to its shadow, 21, as the man, 6, is to his shadow, $4\frac{1}{2}$.

The well is a trifle under 22 feet deep. If the drum is 7 inches in diameter its circumference is 7π and π is about 3.1416. A rope that wraps 12 times around it is $12 \times 7 \times 3.1416$ inches long; cancel the factor 12 and you have the answer in feet.

The answer to the problem of the sheep is not 60, but 55 sheep. If it takes 10 sheep 10 minutes to jump over a fence—the time being measured from the jump of the first sheep to the jump of the 10th—then the interval between jumps is 10/9 of a minute. There are $60 \div 10/9$ or 54 such intervals in an hour, so that 55 sheep jump the fence in this time.

2. DOMINO SETS. In the set up to double-twelve there are 91 bones. The number in any set is the sum of the integers $1+2+3 \ldots +n$, where n is the number of suits. Don't forget that "blank" is a suit. Thus, in the set up to double-six there are seven suits: 0, 1, 2, 3, 4, 5, 6.

3. MARK-DOWN. The selling price was $15.36. The dealer each time deducted 20% of the previous price.

4. NINE DOTS. The solution is shown in the diagram. Hasty readers are apt to reject this solution, once they have found it, because the horizontal line if extended would cross the dot in

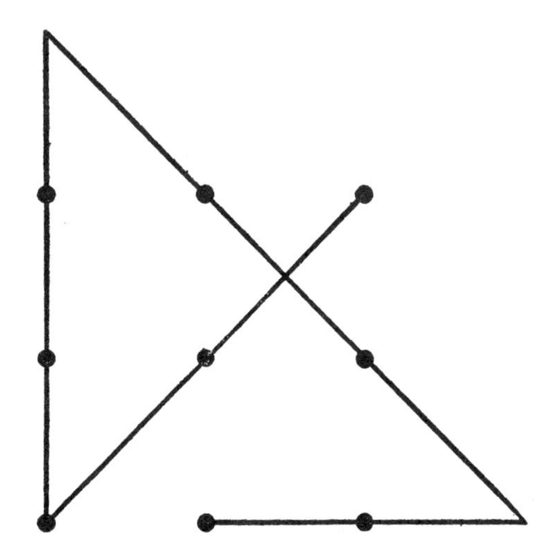

the lower left corner a second time. But nothing in the statement of the problem implies that the line must be construed as infinite in length. On the contrary, the stipulation that the pencil must not be lifted implies that we are dealing with finite line segments.

5. MAKING A CHAIN. The cost is $1.40. The most economical plan is to cut open all four links of one section and use these four to join the five remaining sections together.

6. THE WILY CHIEF. The M'gmb works 14 days per month, earning 70 bmgs. But he loafs on the other 10 days, and is docked just 70 bmgs.

7. THE BOOKWORM. If your answer was 2¼ inches you fell into the trap. When two volumes are in order (left to right) on a bookshelf, the first page of Volume I and the last page of Volume II are separated only by two covers. The correct answer is ¼ inch.

8. AN EASY MAGIC SQUARE.

4	9	2
3	5	7
8	1	6

9. THE FACETIOUS YOUNG MAN. We know that the total of the purchases must be a number divisible by 4 and that it is less than a dollar. The only pairs of digits that sum to 14 are 9, 5 and 8, 6 and 7, 7. The only integer divisible by 4 that can be formed from any pair is 68. Hence the pack of Fumeroles cost 17 cents and the Sure-Fire lighter 51 cents.

10. TANKTOWN TRIOS. Visualize the three outfielders placed at their separate tables. Then a baseman, say first base, may sit at any of 3 tables, and after he is placed the second baseman may sit at either of the 2 other tables. After that there is only one place left for the third baseman. For the first six men there are thus $3 \times 2 = 6$ different arrangements. The pitcher, catcher, shortstop can be seated in 6 ways, so that the total for the nine is $6 \times 6 = 36$. (The same trio will convene 4 times, but on each occasion there will be a variation in the composition of the other two tables.)

11. WATER, GAS, AND ELECTRICITY. The only solution is to persuade one of the house-owners to permit a conduit servicing another house to be laid underneath his own, as shown in the diagram on page 142. Naturally the statement of the problem must be carefully made so as not to exclude this possibility, else the solution when demonstrated will raise a justified protest.

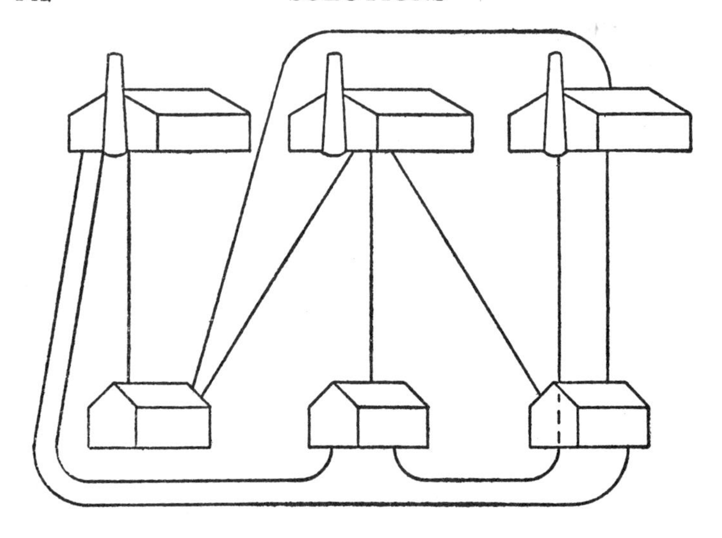

12. A BRICKEY QUICKIE. A whole brick weighs 3 pounds.

13. SPOTTING THE COUNTERFEIT. Divide the coins at random into three groups of three. Balance any two groups against each other; if one contains the underweight coin the group is spotted; if the two groups balance, the counterfeit is in the third group. From the spotted group take any two coins and balance them. If one is light, it is the counterfeit; if they balance, the counterfeit is the third of the group, left on the table.

14. THE PAINTED CUBE. Only one of the 27 small cubes is unpainted; 8 are painted on three faces, 12 on two faces, 6 on one face. These numbers correspond to the number of corners, edges, and faces of the large cube.

15. SHEEP AND GOATS. The solutions are given by naming the pens from which the pair is to be moved; it is unnecessary to name the pens to which they are transferred as only two pens are vacant.

The first puzzle is solved by moving 4—5, 6—7, 2—3, 7—8. The second is solved by 3—4, 6—7, 1—2, 7—8.

16. THE BILLIARD HANDICAP. Huntingdon should give McClintock 40 points in 100. If he gives Chadwick 20 in 100, then Chadwick is expected to make 80 while Huntingdon makes 100, thus scoring points at $\frac{4}{5}$ of Huntingdon's rate. Similarly, McClintock scores at $\frac{3}{4}$ the rate of Chadwick. Then McClintock should score at $\frac{4}{5} \times \frac{3}{4}$ or $\frac{3}{5}$ the rate of Huntingdon.

17. THE SURROGATE'S DILEMMA. William Weston's "evident intention" was to divide the estate 2:1 as between son and widow, or 1:3 as between daughter and widow. These ratios can be preserved by awarding the son $\frac{6}{10}$ of the estate, the widow $\frac{3}{10}$, and the daughter $\frac{1}{10}$.

18. THE LICENSE PLATE. If the plate showed a readable number upside-down as well as right-side-up, then all the digits on it were reversible. Only five of the digits can be written so as to be reversible—1, 6, 8, 9, 0. Since all five digits on the plate were different, they must have comprised just these five. The problem then is to arrange the digits to make an integer which is 78,633 less than its inversion. A little trial will discover the number to be 10968. A point to be remembered is that 1, 8, 0 stay the same on inversion, but 6 and 9 exchange identities.

19. MEASURING TWO GALLONS. Fill the 5-gallon can and then pour its contents into the 8-gallon measure. Fill the 5-gallon can a second time, and then fill the 8-gallon measure from it. As the latter will take only 3 more gallons, 2 gallons are left in the can.

20. MATCHSTICK EQUATIONS. The illustration shows the form of each equation after the position of one match has been changed.

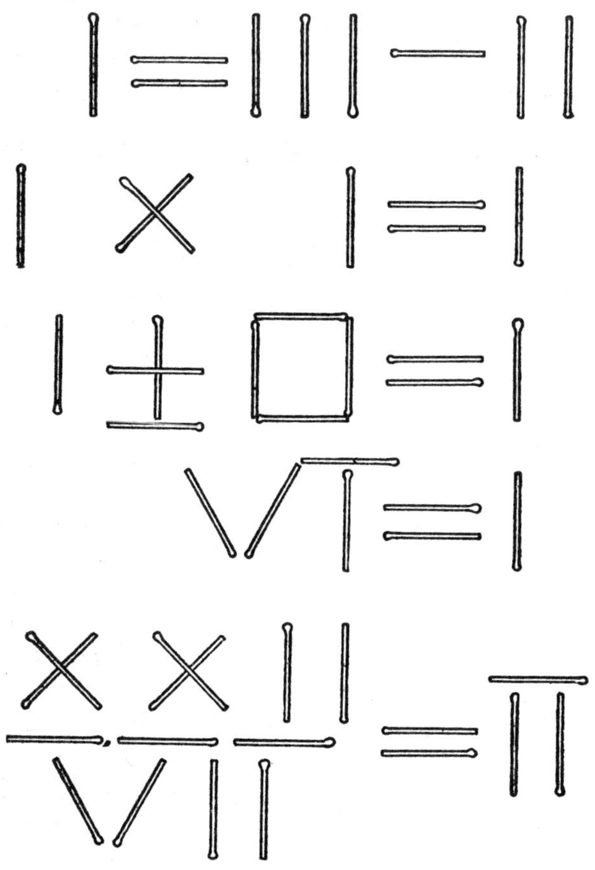

In the first, the transfer of one match causes the minus sign and the equals sign to exchange places.

In the second, the minus sign is removed and made into figure 1, while the Roman 10 now becomes a multiplication sign.

In the third, an intrusive match is added to the plus sign to make it read "plus-or-minus."

In the fourth, the match moved is used to transform the Roman 5 into the radical sign meaning "the square root of."

In the last, the extra match changes Roman 2 into Greek π, and if your victim argues that $\pi = 3.1416$, which is not equal to $\frac{22}{7}$, you can retort that the latter value was accepted as correct in archaic times and still is used as a close approximation.

21. WHAT IS THE NAME OF THE ENGINEER? The businessman who lives nearest the engineer is named Smith, and the engineer's income is exactly one-third of his. This businessman cannot be the brakeman's namesake, for the latter is said to earn $3500 and this number is not divisible by 3. (If that point in the argument seems thin, don't argue with *me!*) Therefore the brakeman's name is not Smith. Nor is the fireman's name Smith, because railwayman Smith beats him at billiards and so must be a different person. Hence the name Smith can be attached only to the engineer, and the question is answered.

22. AT THE RAINBOW CLUB. White was not highest, as his card was lower than another. Black was not highest, as his partner had a choice. Brown was not highest, since he was right-handed, and the choice of cards fell to a left-handed man. Hence, Green drew highest card. Green was not White's partner, and Brown was not White's partner. Therefore Green and Brown were partners against White and Black, with Brown at White's left. Black drew the lowest card, since White had the choice for his side.

23. TENNIS AT HILLCREST. The minimum number of sets that could have decided the tournament was 15, totaling 90 games for the winners. The winners actually took 97 games (4). One extra set was played in the first round (3), leaving 1 game to be accounted for. One set in the whole tournament must have reached 5-all and must have been won at 7—5.

Bancroft lost his first match by 6—4 and 7—5 (7).

Franklin reached the final, where he lost (8). Since he won the unique 7—5 set, his first-round opponent was Bancroft.

Other first-round pairings were Abercrombie vs. Devereux (5), and Gormley vs. Eggleston (9). The remaining two entrants must have been paired: Haverford vs. Chadwick.

The winners in the first round were Haverford (3), Franklin (8), Eggleston and Devereux (6).

In the second round Eggleston did not meet Haverford (1), nor did he meet Franklin, for Franklin vs. Bancroft and Eggleston vs. Gormley were in different halves of the original bracket (2). Therefore Eggleston met Devereux, and Haverford met Franklin. The winners were Devereux (6) and Franklin (8).

Devereux won the final from Franklin by 6—4, 6—4, 6—4. (8).

24. WHITE HATS AND BLACK HATS. If one man, say A, wore a white hat, then B would know that he himself must have a black hat, else C would not have raised his hand. By the same token, C could infer that his own hat was black because B's hand was raised. When neither B nor C spoke up—the nugget of the narrative is the word "presently"—A knew that his own hat could not be white.

25. TRUTH AND FALSEHOOD. Every inhabitant was bound to say that he was a Diogene—the Diogenes because they were truthful and the Ananias because they were liars. Hence the second man's assertion must have been false, and since the third spoke truly he was a Diogene.

26. WINE AND WATER. There is exactly as much water in the demijohn as there is wine in the bucket. Regardless of the proportions of wine and water transferred—and regardless of the number of exchanges—if the two containers first hold equal volumes of pure liquid and eventually are left with equal volumes of mixtures, equal amounts of wine and water have changed places.

27. FOUR PENNIES. This is something of a "catch" problem. Manifestly there is no solution if the pennies must be kept in one plane. The trick is to arrange three pennies in a triangle and put the fourth on top of any one of the others, creating in effect a "double point."

28. SEVEN PENNIES. Each penny after it is moved along a line must come to rest *adjacent* to the penny or pennies already placed. The adjacency of points of course refers to the line connections, not to the circle on which the points lie. Every added penny must therefore begin two points away from one already down and must move to the intervening point.

29. THE ROSETTE. The rosette contains seven coins, including the one in the center. Perhaps you remember from geometry that a true hexagon, a regular figure with six sides, is inscribed in a circle by laying out six chords exactly equal to the radius of the circle.

30. THE MISSING PENNY. It would be correct to sell 3 apples of the inferior grade and 2 of the superior, together, for twopence. Any amount of the combined stock may properly be sold at 5 apples for twopence, provided that the ratio of 3:2 is maintained. The sale of 60 apples would have come out right had there been 36 of the cheaper kind and 24 of the other. But with 30 of each, 6 of the better apples were sold at the cheaper rate, for a loss of 6 times the difference between the rates.

31. THE RUBBER CHECK. The dealer gave the customer a radio and some cash in exchange for a worthless piece of paper. His loss was $43.75 plus $10.02, a total of $53.77.

Of every three solvers, two will probably argue that this answer is wrong. One will say that the dealer also lost $26.23 profit on the radio given to the customer. But this is not cash out of pocket and would not be entered as loss on the books. The other solver will argue that from the loss of \$53.77 must be deducted \$26.23 profit made on the radio given to the landlord. But why is this particular item deductible? Why not, then, all the profits made by the dealer on all his sales? There is nothing in the problem to indicate that the landlord would not have purchased a radio in any case, or that the radio, if not sold to the landlord, could not have been sold to someone else.

32. MYSTERIOUS COMPUTATION. The professor told Edward that the computations were made in the heptary system (radix 7) instead of the denary system (radix 10).

When we write a juxtaposition of digits such as . . . CBA, in the denary system we mean the cardinal number

$$A(10^0)+B(10^1)+C(10^2) \cdots$$

If we use a system based on 7 instead of 10, the same digits will express a different number, namely

$$A(7^0)+B(7^1)+C(7^2) \cdots$$

A formula like the above serves to convert the expression of a number from one radix to another. If all numbers on the scrap of paper are translated into the more familiar denary notation, they are seen to be a correct division of 999 by 37 to give quotient 27, and the sum

$$\begin{array}{r} 92 \\ 74 \\ 82 \\ \hline 248 \end{array}$$

The reader may think that Professor Digit dipped into the heptary system merely to puzzle his son. But the fact is that certain aspects of mathematical theory become clearer if a system of notation other than the denary is used. For example, games like No. 178—*Three-Five-Seven* require the conversion of numbers to the binary scale (radix 2).

33. THE TENNIS TOURNAMENT. Perhaps you laboriously set up the brackets, as I did when asked this question. I felt foolish when it was pointed out that 77 of the entrants have to be eliminated, hence 77 matches are required.

34. TARTAGLIA'S RIDDLE. Four. Set up the proportion

$$\frac{5}{2} : 3 = \frac{10}{3} : x$$

and solve for x. The argument is: Whatever mysterious factor causes $\frac{1}{2} \times 5$ to give the result 3 must also be introduced into the product $\frac{1}{3} \times 10$. This factor is expressed by the ratio $5/2:3$.

35. STRANGE SILHOUETTES. The object is a type of paper drinking cup dispensed in American railway cars. The cup folds flat, in which condition its shape is the frustum of a cone.

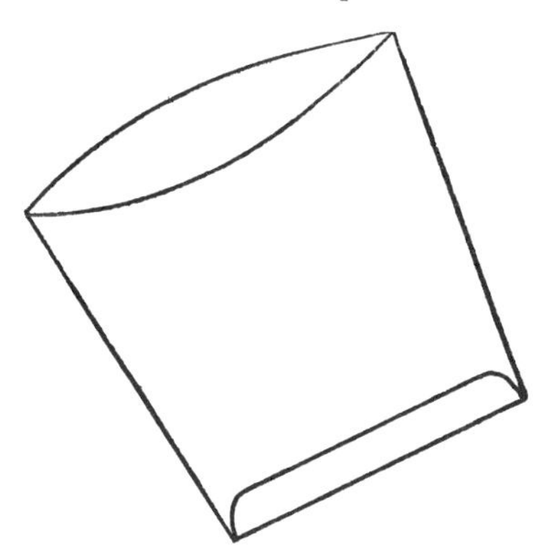

When it is opened out, with the mouth made circular, the creases at the side become parallel. It might then be described as a kind of cone with a circular base whose elements do not meet in a single point but which all intersect a line-segment parallel to the base and equal to its diameter.

This puzzle was invented, however, long before the paper drinking cup!

36. THE DRAFTSMAN'S PUZZLE. A consistent side view is shown in Fig. 1, and Fig. 2 is an oblique view of a block that

Fig. 1 Fig. 2

satisfies the drawings. The broken lines can be taken to indicate a rectilinear lug projecting from the surface, or a cavity gouged out of the block.

37. A PROBLEM IN PROBABILITIES. Three. If you were caught on this question and gave some answer that involved 10 and 16, you are not the first!

38. A LAMICED PUZZLE. Rallod, enough for anyone!

39. BEAR FACTS. The bear was white. The last lap of his tour is at right angles to the first leg. If the bear is headed south on reaching his den, and left it in a due south line, his den must be on the North Pole, from which every direction is due south. Hence he is a polar bear.

40. THE FLAG OF EQUATRIA. The agreement of the Equatrian states on the principle of dissection suggests that the tessellation shows how the six-pointed star may be dissected to useful purpose. By experiment we can discover that the pieces of each star can be rearranged to form a single equilateral triangle, the emblem of the Equatrian union. The Punroe Doctrine proceeds from the fact that these five stars represent the only ways in which a six-pointed star can be dissected into no more than five pieces for this purpose. An interesting collateral inquiry is to prove that no other solutions exist.

Hint: The side of the equivalent triangle is the distance between opposite vertexes. In any five-piece solution, all pieces must be compounded out of a basic unit, which is the small shaded triangle. The star is composed of 12 such triangles. There are 13 partitions of the integer 12 into five parts. Investigate which partitions can give geometrical solutions and which cannot. I have considered a solution to be defined by the size and shape of the component pieces, not by the ways in which they can be assembled to form either the star or the triangle. In other words, I do not count, as separate solutions, different ways of making cuts to get the same five pieces.

42. A QUESTION OF BARTER. For a fishhook, 2 coconuts; for a knife, 4; for a spear, 5.

43. SHARING APPLES. The boys gathered 72 apples.

44. A TRANSACTION IN REAL ESTATE. The loss was $200. The tax bill was $500 and the repairs cost $1,000.

45. SETTLING THE BILL. Originally there were 8 men in the party.

46. COWS AND CHICKENS. The wight had 4 cows and 31 chickens.

47. THE FARMER'S RETORT. One cow, 34 chickens, and 2 stools.

48. DOLLARS AND CENTS. Blake arrived with $12.35, having started with $35.12.

49. THE JAY ESTATE. The estate was $40,000.00. Note that "30 times more" means "31 times as much."

50. A FISH STORY. The mackerel was 21 inches in length: head 4, body 12, tail 5. The pickerel's measurements were: head 3, body 8, tail 4.

51. WHO NOES? NOT AYE! The total number of voters was 60. The first vote was 36—24 against the motion. The second vote was 30—30. The third was 31—29 against.

52. NO FREEZEOUT. The first loser must have had $4.05 before the last round; the second, $2.05; the third, $1.05; the fourth, 55¢; the last, 30¢. This problem is easy enough to solve by working backwards. Perhaps the reader discovered the formula whereby the problem can be solved for any number of players and any final amount. This final amount is of form $m(2^n)$ where n is the number of players. The last loser must have started with $m(n+1)$, the next-to-last with $m(2n+1)$, the third-last with $m(4n+1)$, and so on to the first loser, who started with $m(2^{n-1}n+1)$. In the present puzzle, $n=5$ and the final amount is $1.60, so that $m=5$.

53. JOHNNY'S INCOME TAX. The gross income was $1064.00. The tax was $106.40.

54. SPENDING A QUARTER. Fourteen sheets of paper (7¢), 8 pens (8¢), 2 pencils (5¢), and 1 eraser (5¢).

55. THE SPOOL OF THREAD. Mrs. Plyneedle had 99¢, in coins of 50¢, 25¢, 10¢, 5¢, and four 1¢ pieces. The thread was 19¢ per spool. The selection of coins was, for one spool: 10, 5, 1, 1, 1, 1; for two spools: 25, 10, 1, 1, 1; for three spools: 50, 5, 1, 1; for four spools: 50, 25, 1.

56. A DEAL IN CANDY. One boy chose 20 chocolate bonbons—notice that it is not stated that each boy took some candy of each kind. Another boy decided on 5 jujubes, 12 bonbons, and 3 lollipops. The third took 10 jujubes, 4 bonbons, and 6 lollipops.

57. WHAT SIZE BET? The bet was 60¢. High hand put in a dime and a half-dollar. Second hand put in a half-dollar and 2 nickels. Third hand put in a dime and a silver dollar, taking back a half-dollar. Fourth hand put in a silver dollar and 2 nickels, taking out the other half-dollar. Last man put in a silver dollar, taking out the 2 dimes and 4 nickels, leaving just 3 silver dollars.

The try that catches some solvers is to make the bet 35¢. The plan is to have the last man chip a dollar, leaving this coin plus a half-dollar and a quarter in the pool and taking the rest. But the pool is cleaned out of quarters before his turn comes.

58. THE HOSKINS FAMILY. There are 6 women, 5 men, and 3 children. From the facts given it can be determined that the quota for a woman is 8 times that for a child, and for a man is 13 times that for a child. The problem is to find integral values for the letters in the equation

$$8W+13M+C=116$$

with the proviso that $W>M>C$.

59. THE TIDE. The force of the tide was ¾ of a mile per hour. All rate problems depend on the formula

$$rt=d$$

where r is the rate, t the time, and d the distance. Let r here stand for the force of the tide, and R for the rate of the boat in still water. Then the net speed of the boat going with the tide is $R+r$ and against the tide is $R-r$.

60. LOCATING THE LOOT. Seven minutes elapsed between the car's two meetings with the trooper. Two minutes were lost in stopping and turning. To reach a point 2 miles from the booth required 1½ minutes. That left 3½ minutes to go beyond to the cache and double back to the same point. In this time the car would travel 4⅔ miles, or 2⅓ miles each way. Hence the car could have got no farther than 4⅓ miles beyond the booth.

61. STRIKING AN AVERAGE. The first 5 miles were covered in 10 minutes. To go 10 miles at the rate of 60 miles per hour requires 10 minutes. The motorist can strike this average only by being transported instantaneously from the halfway mark to his destination.

62. THE SWIMMING POOL. Jill won the race. The pool was 42 feet long.

63. HANDICAP RACING. The little boy overtook the little girl 15 seconds after the start of the race, or 5 seconds after the big boy passed her.

64. THE PATROL. The first occasion when all three patrolmen could meet at point A was an hour and a half after leaving it simultaneously. Therefore the radio message arrived at 4:30 A.M.

65. THE ESCALATOR. As this problem has proved puzzling to many, I will show how to attack it. Let x be the number of steps from bottom to top. Let t be the time required from any one step to displace the one immediately above it. A person standing still would require a time xt to go from bottom to top. But Henry walked up 28 steps and so reached the top on that step which was $x-28$ from the top when he started at the bottom. The time of his trip was therefore $(x-28)t$. Since he took 28 steps in this time, he walked at the rate of $\frac{(x-28)t}{28}$ per step, or two steps in the time $\frac{(x-28)t}{14}$. In the same time Martha took one step, and by the same form of computation her time per step was $\frac{(x-21)t}{21}$. Equate the two latter terms and solve for x. The answer is that the escalator was 42 steps long.

66. THE CAMPER AND THE BOTTLE. The camper reached a point $3\frac{2}{3}$ miles upstream from his camp. He first met the bottle 2 miles away—the distance he could travel in 48 minutes at the net rate of $2\frac{1}{2}$ miles per hour. The bottle then floated 2 miles at the rate of the stream, taking 80 minutes to reach the camp. In this time the canoer, with respect to the water, went upstream some distance and then returned to his starting point (the distances and times of the two trips must be equal). The total time being 80 minutes, the upstream trip occupied 40 minutes. With respect to the banks of the stream, the canoer still proceeded at the net rate of $2\frac{1}{2}$ miles per hour, and therefore traveled an additional $1\frac{2}{3}$ miles away from his camp.

67. HITCH AND HIKE. The trip was done in 5 hours and 6 minutes. By jeep alone the transport would have required 7 hours and 30 minutes. The hiking therefore saved 2 hours and 24 minutes.

The arrival of the parties simultaneously might have been the consequence of a decision on the part of the sergeant to make each party walk just 16 miles. He dropped the first party 16 miles from the rendezvous, picked up the second after it had walked 8 miles and took it to a point 8 miles from the rendezvous, and picked up the last party after it had walked 16 miles.

68. IF A MAN CAN DO A JOB. Working alone, the tinker would take 3 days and the helper would take 6. The apprentice would never finish the job at all—in fact, he turned out to be entirely useless.

69. FINISH THE PICTURE. One cube and one sphere. The relative weights of the solids are: cylinder 13, cube 8, sphere 4, cone 3.

70. THE ALCAN HIGHWAY. Fifteen weeks behind schedule. That is, the whole job will take one bulldozer sixteen weeks.

71. SEE SAW. Alfred must sit $37\frac{1}{2}$ inches from the center point. He weighs 32 pounds and Bobby weighs 60 pounds.

72. A PROBLEM IN COUNTERWEIGHTS. Obviously a 10-pound weight is needed. Then a 20-pound weight will make it possible to offset a load of 20 or 30 pounds, the latter by addition of the 10-pound weight. Next a 40-pound weight will make possible compounds of weights of 40, 50, 60, 70 pounds. Evidently the series of the most economical weights is

$$10 \times 2^n$$

where n takes all integral values 0, 1, 2, etc. The five counterweights are therefore 10, 20, 40, 80, 160, reaching a total of 310 pounds.

73. THE APOTHECARY'S WEIGHTS. This problem differs from *A Problem in Counterweights* because here we can use subtraction as well as addition.

It will perhaps be clearer if we deal with units of 1 instead of ½. After we have determined the proper set to weigh out all integral quantities from 1 up, we can divide the several weights by 2.

We have to start with a weight 1. To measure 2, it is most economical to add a weight 3, for then we get 2 by subtraction and can also make 3 and 4. To get 5 *et seq.* we can make the third weight 9, for then subtraction of the 4 and the lesser integers gives 5, 6, 7, 8, while the 9 alone and additions of 1 to 4 give 9, 10, 11, 12, 13. It is evident that the most economical series of weights is given by 3^n, where n takes all values from 0, 1, 2, 3, up. The first five weights in the series are 1, 3, 9, 27, 81. Divide by 2 for application to the problem: Weights in grams of ½, 1½, 4½, 13½, and 40½ reach the maximum possible total of 60½.

74. SALLY'S AGE. Sally's age is 22 years and 8 months.

75. AS OLD AS ABC. Alice is 8; Betty is 5; Christine is 3.

76. FUMER FROWNS. Because the shop actually lost 10¢ on the sale. The Vesuvius pipe must have cost $1.50, and the 20% loss was 30¢. The Popocatepetl pipe cost just a dollar, and the 20% profit was 20¢.

77. COMPOUND INTEREST. Like many algebraic problems, this is easiest solved by determining the general formula applicable to any rate of interest and any number of compoundings, then filling in the actual values given in the problem.

Let a be the initial deposit and r be the rate. Then the interest at first compounding is ar and the total is $a+ar$. Interest at the second compounding is $(a+ar)r$, and the total is this quantity plus $(a+ar)$. The whole expression reduces to

$$a(1+2r+r^2)$$

By continuing the process through further compoundings, it will be seen that the total s is always equal to

$$s=a(1+r)^n$$

where n is the number of times interest has been compounded.

Applying the formula to the problem, we have

$$\$100.00=a(1+.03)^5$$

The expansion of the right-hand member looks more formidable than it is. Treat it as a binomial and apply the binomial theorem. The reader who has forgotten the theorem can look up the coefficients in "Pascal's triangle" under No. 136—*Figurate Numbers*. These coefficients are 1, 5, 10, 10, 5, 1. If the terms of $(1+.03)^5$ are written out, without expansion, it can be seen that they decrease in magnitude very rapidly. If we take only the first three terms $(1+.15+.009)$ and accept the sum as about 1.16, we shall be near enough for practicable purposes. With this value in the equation, we get

$$a=\$86.20$$

which is accurate within the statement of the problem. The actual sum which must be deposited is $86.25.

78. THE SAVINGS ACCOUNT. The general formula for compound interest is given in the solution above. In the present problem we have

$$\$131.67=(\$100.00)(1+r)^8$$

The solution of this equation boils down to finding the 8th root of 1.3167. This can be done by extracting the square root three times. If the operations are carried to only four places of decimals, r is found to be .0351. The rate is actually $3\frac{1}{2}\%$.

79. AFTER FIVE O'CLOCK. These puzzles, like many others concerning clocks, are based on the circumstance that the minute hand travels 12 times as fast as the hour hand. The basic equation for such puzzles is

$$D = 12d$$

where D is the distance traversed by the minute hand and d the distance traversed by the hour hand.

In the ordinary garden variety of clock, the hands move by little jerks. They stand still for an interval of one or two seconds, or even, in many electric clocks, for a full minute. But puzzle books are inhabited by a very special kind of clock in which the hands move continuously.

(a) $27\frac{3}{11}$ minutes past five. The second equation here is $D = d + 25$.

(b) $10\frac{10}{11}$ minutes past five. At right angles, the hands are 15 minutes apart. Hence $D + 15 = d + 25$.

(c) $16\frac{4}{11}$ minutes. The involved statement of the question is equivalent to asking: What is the interval of time between coincidence of the hands and the next position 90 degrees apart? The answer can be found simply by subtracting the answer to (b) from the answer to (a).

80. THE CARELESS JEWELER. The watch first shows the correct time at $5\frac{5}{11}$ minutes past three. During the hour from two to three the watch is continuously wrong. The minute hand moves from XII to I while the hour hand makes a complete circuit from II to II. Between three and four o'clock, the hour hand makes a second circuit, and the watch will be correct at one instant while it is in transit between III and IIII. Let d be the distance from II to this point. Let D be the distance the minute hand advances beyond I during the same time. Because of the reversal of the hands

$$d = 12D$$

When the same position is reached by a correct watch, the hour hand moves forward from III a distance d—5, while the minute hand travels from XII a distance D+5, and this distance is 12 times that moved by the hour hand:

$$D+5=12(d-5)$$

The answer to the problem is found by solving the two equations simultaneously.

81. CLOCK SEMAPHORE. (a) At 18 6/13 minutes past two. (b) At 55 5/13 minutes past twelve.

83. NOW REVERSE IT. From the discussion of No. 82—*Changing a Rectangle to a Square* the reader may have jumped to the conclusion that a rectangle of any dimensions can be dissected to make an equal square by *only three pieces*. Such a conclusion is incorrect.

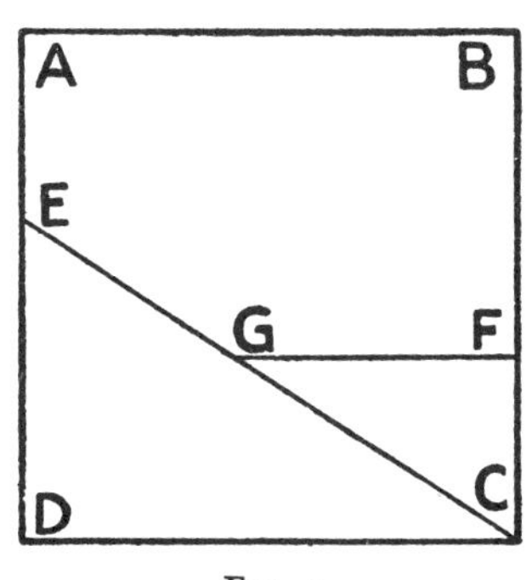

FIG. 1

Let us look at the question in the reverse way. Given the square ABCD in Fig. 1, how elongated a rectangle can we make by the three-piece dissection? If ED is to be the height of the rectangle, connect EC and cut. On BC measure BF equal to ED, through F draw FG parallel to AB, and cut on this line. Slide the upper piece down to the right so that its corner G coincides with C. Transfer the small triangle so that FC coincides with AE.

It is evident that when ED is just half of AD, the point G coincides with E, and FE divides the square into two equal rectangles. We can then omit the cut EC, and by cutting into only two pieces produce a rectangle whose sides are in ratio 1:4.

It is also evident that if ED is less than half of AD, the three-piece cutting will not work. We will have to dissect into more pieces in order to produce a rectangle more elongated than 1:4. Conversely, a rectangle whose sides are in greater ratio cannot be cut into only three pieces to form a square.

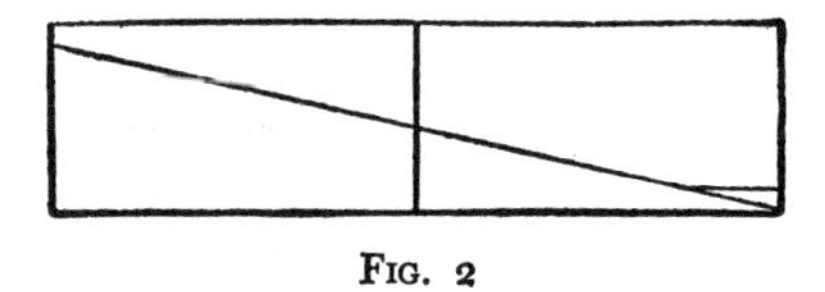

FIG. 2

The line B given in *Now Reverse It* is in length a little more than twice the side of the square A. The solution is to cut the square into two equal rectangles, set them side by side as in Fig. 2 to form a rectangle 1:4, then make the three-piece cuts to form the more elongated rectangle of side B. This divides the original square into five pieces.

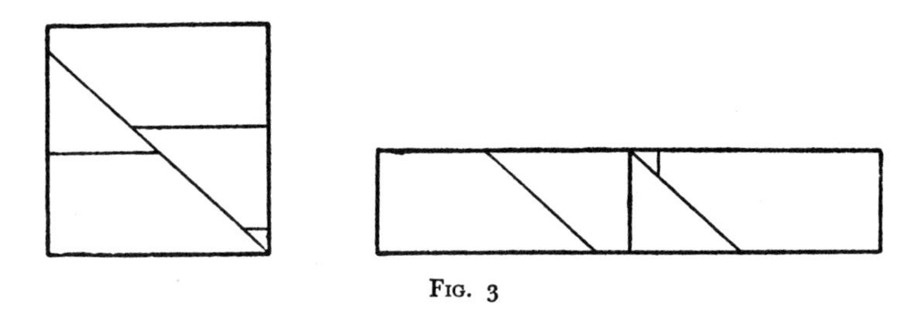

FIG. 3

Another solution simply reverses the order of the two steps. First cut the square into three pieces to form a rectangle whose longer side is half of length B. Then cut this rectangle into two equal rectangles which can be set side by side to form the final figure. Fig. 3 shows that again we must make five pieces.

In all the above, we have fixed the position of cuts by measurement with the *smaller* side of the desired rectangle. The given line B is the *larger* side. In case any reader is puzzled how to determine the smaller side, I give the construction in Fig. 4. AB is the given length B, while BC, perpendicular to it, is the side of the

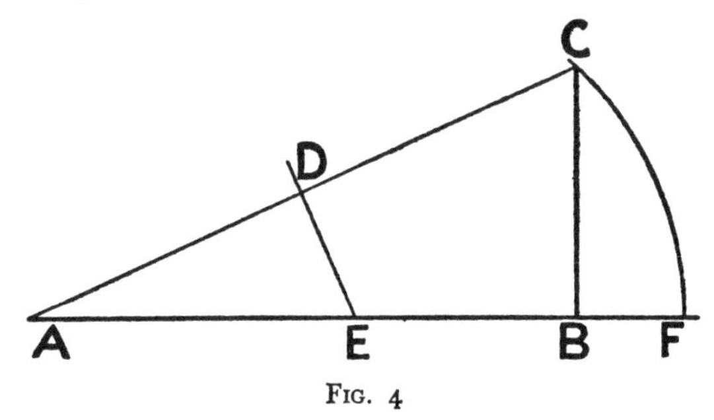

FIG. 4

square A. Connect AC. Construct the perpendicular bisector DE, of AC, cutting AB in E. Then AE equals EC; measure this same length from E toward B to fix point F. The desired length is BF. [Since AB:BC=BC:BF, or $(AB)(BF)=(BC)^2$.]

84. THE KITCHEN LINOLEUM. The linoleum need be cut into only two pieces, as shown by the diagram. Each "step" is 4 feet wide and 3 feet deep. If the right-hand piece is moved one step down and leftward, a square 12 by 12 is formed.

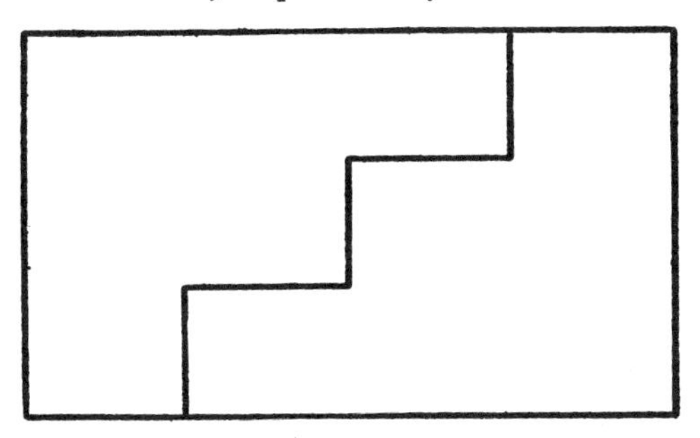

This "step" principle is the basis of many rectangle-to-square puzzles, but there is more to it than meets the eye!

85. THE BROOM CLOSET. The diagrams show how to cut

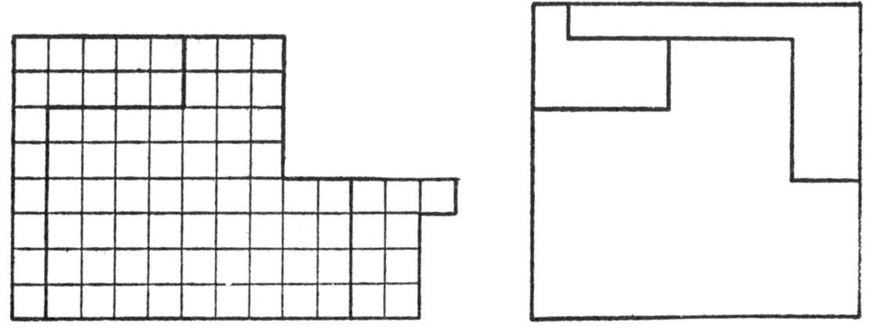

the linoleum into 3 pieces that can be fitted together to form a square, correct in pattern.

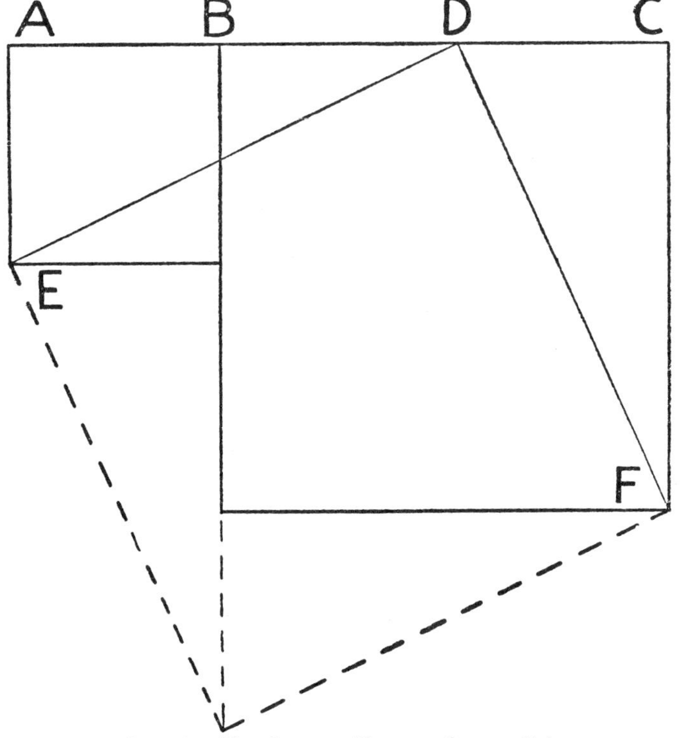

Summing Two Squares (See top of page 163)

86. SUMMING TWO SQUARES. Set the two squares side by side as shown in the diagram. Measure off CD equal to AB, the side of the smaller square. Connect DE: this is the side of the total square. Connect DF: this marks a second side equal to DE. Cut along DF and DE. Bring the 3 upper pieces down to the positions shown by the broken lines to form the total square. This demonstration shows that *any* two different squares can be dissected into one through not more than 5 pieces.

87. FROM A TO Z. The diagrams show how to cut the A into 4 parts to make a Z. The secret of avoiding having to turn over a piece lies in that little equilateral triangle.

88. THE MITRE FALLACY. The mitre indeed can be made into a rectangle, but this rectangle cannot be cut into a square on the "step" principle.

To make the proof concrete, let us say that the mitre is made out of a square 4 units on each side. (We can choose any number we please, since we will deal with ratios, and ratios are the same whatever the unit of measurement.)

Then the rectangle made from the mitre is 3 by 4.

The width of all steps must be the same, and the heights must be the same. Hence, with 7 widths along the 4-unit edge and 6 heights along the 3-unit edge, each step is $4/7$ wide and $1/2$ high.

Sliding the right-hand piece down one step adds $1/2$ to the depth of the rectangle and subtracts $4/7$ from its width. The resultant "square" is $3\frac{3}{7}$ wide by $3\frac{1}{2}$ deep. That is, it is not really a square.

Nor can a different number of steps be found that will yield a true square. The simplest way to prove this fact is to investigate the conditions necessary so that a rectangle can be cut into two pieces stepwise to form a square.

Let the sides of the rectangle be a and b, with a the greater. Let the dimensions of each step be w and h, w being parallel to a. Let $a/w=n$, the number of steps counted parallel to a. Let $b/h=n'$, the number of steps parallel to b.

Manifestly, the dissection is feasible only if n and n' are both integral and $n=n'+1$.

Furthermore, the dimensions of the altered rectangle will be $a-w$ and $b+h$; the figure is a square only if these two are equal (and consequently both equal to $\sqrt{ab}$.).

From the foregoing equalities we can express n and n' in terms of a and b:

$$n=\frac{a}{a-\sqrt{ab}} \text{ and } n'=\frac{b}{\sqrt{ab}-b}$$

A rectangle yields to stepwise dissection only if its dimensions give integral values for n and n' in the above equations, such that $n=n'+1$.

Loyd's proposed solution of the mitre problem deals with a rectangle whose sides are in ratio 3:4. These values do not satisfy the above conditions.

89. WHAT PROPORTIONS? From the equations given under *The Mitre Fallacy* the following can be derived:

$$\frac{a}{b}=\frac{n^2}{(n-1)^2}$$

The present problem gives $n=13$. Then, the two sides of the rectangle must stand in the ratio $13^2:12^2$, or 169:144.

The above equation serves to point out that a rectangle yields to stepwise dissection into a square only if its two dimensions stand in the ratio of two consecutive integral squares.

90. THE ODIC FORCE. The diagrams show the dissection. The bottom third of the large square is cut off and this rectangle

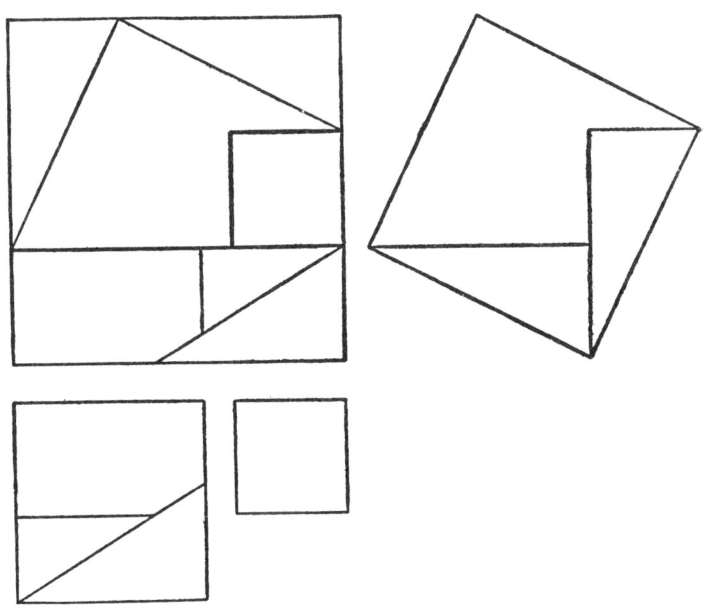

is cut into 3 pieces to make a square by the method explained in No. 82—*Changing a Rectangle to a Square*. This is the square of area 3. The square of area 1 is cut out intact from a corner of the remaining piece, its side being ⅓ the side of the original square. The third piece, of irregular shape, can be cut into 3 pieces to make the square of area 5. The point on the base where the two cuts meet is ⅓ the distance from left to right.

91. THE PIANO LAMP. The diagrams on page 166 show how the piano lamp can be cut into 10 pieces that can be fitted together to form a circular disk.

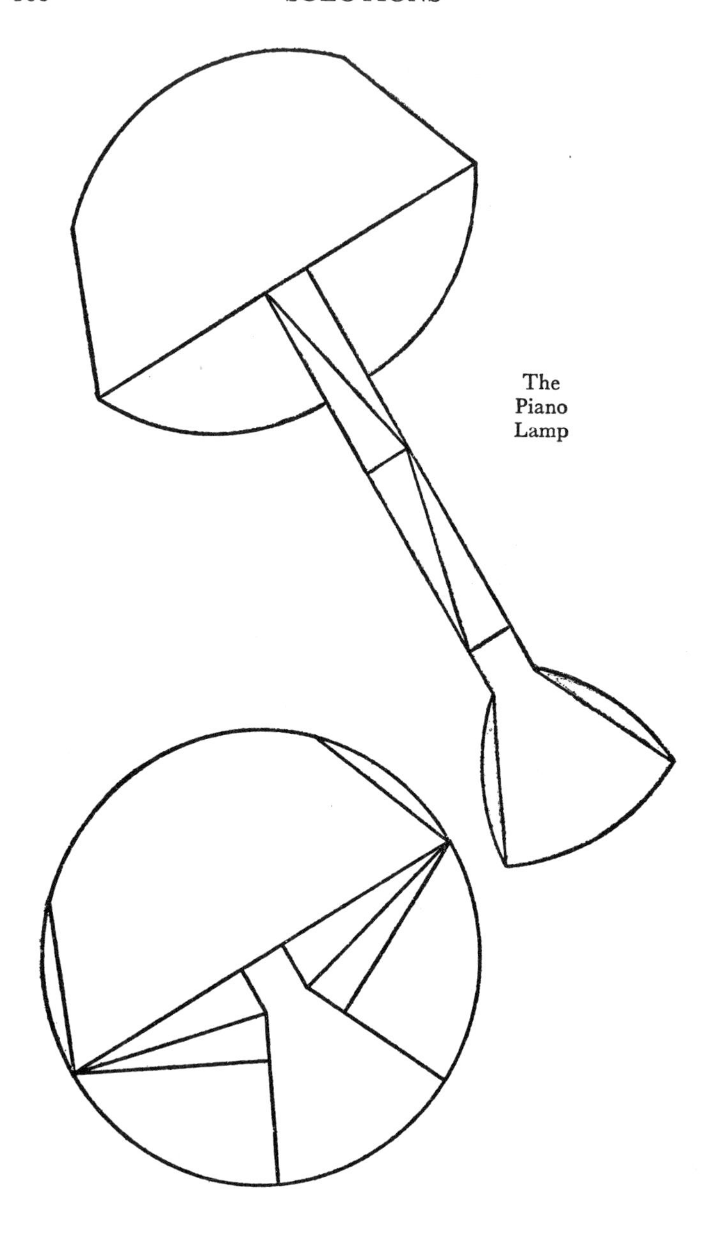

The Piano Lamp

92. CONSTRUCTION OF A PENTAGON. In the diagram given with the construction, BC is the hypotenuse of a right triangle OBC. Hence $BC^2=CO^2+OB^2$. We can treat the circle as a unit circle and set OB equal to 1. And CO=AC—AO; AO=½; AC=AB=$\sqrt{5}/2$. From these figures BC is found to be 1.17558.

(The correctness of the construction can be verified by trigonometry. The side a of the pentagon inscribed in a circle of radius r is given by $a=2r \sin 36°$. Sin 36°=.58779, and for $r=1$, $a=1.17558$.)

93. THE AMULET. The dissection of a pentagon into a square in only 6 pieces was for some time a baffling problem, and its solution by Dudeney deserves to be better known than it is.

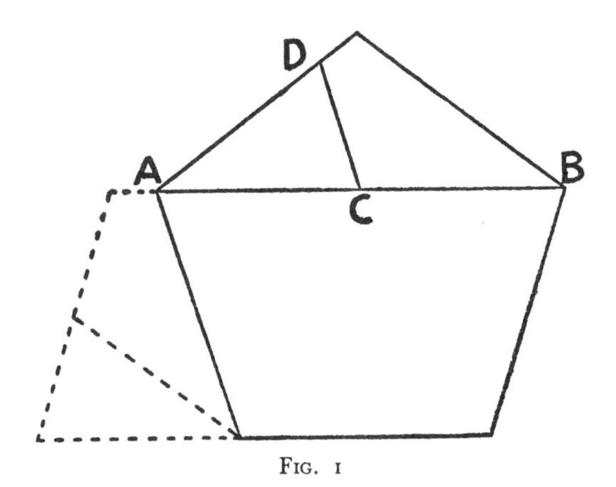

FIG. 1

Fig. 1 shows how the pentagon is cut into 3 pieces to make a parallelogram. One cut is on the line AB and the other on CD; C is the midpoint of AB and AD is made equal to AC. The broken lines show how the 2 upper pieces are rearranged around the lower piece.

In Fig. 2 the parallelogram is dissected by two cuts (broken lines) to make a square. The distance FB is the mean proportional between the base EB and the height of the parallelogram. (How

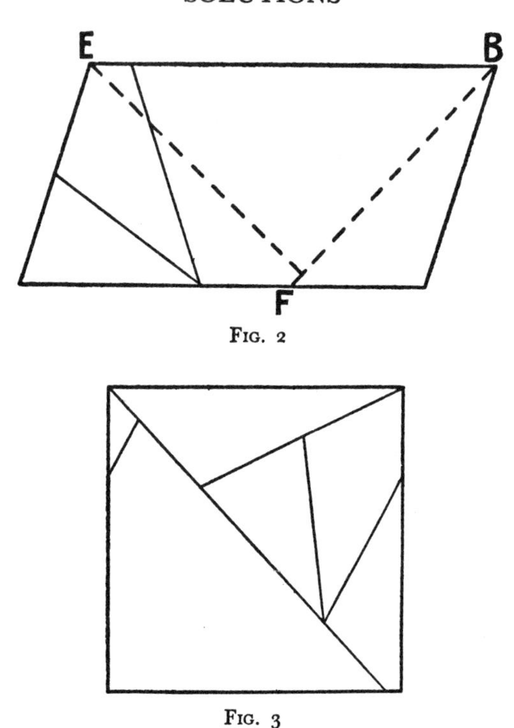

FIG. 2

FIG. 3

to find the mean proportional is described under No. 82—*Changing a Rectangle to a Square.*) The two cuts divide the parallelogram into 6 pieces altogether, which make a square as shown in Fig. 3.

94. FOUR-SQUARE. There are 10 different pieces, as shown by the diagram.

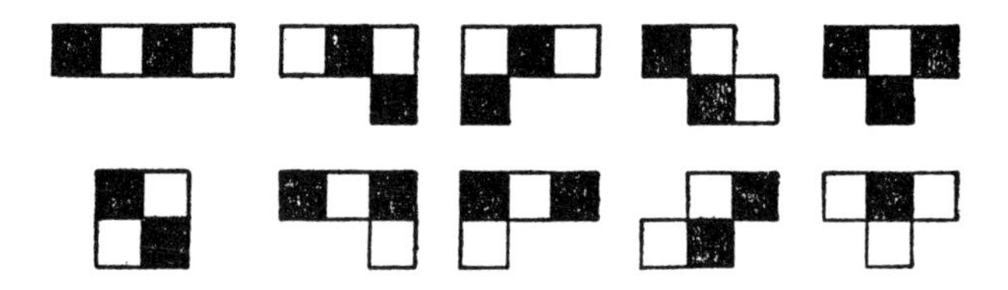

95. JACK O' LANTERN. The diagrams (Figs. 1 and 2 on pages 169 and 170) show how to cut the disk and arrange the pieces to form the jack o' lantern.

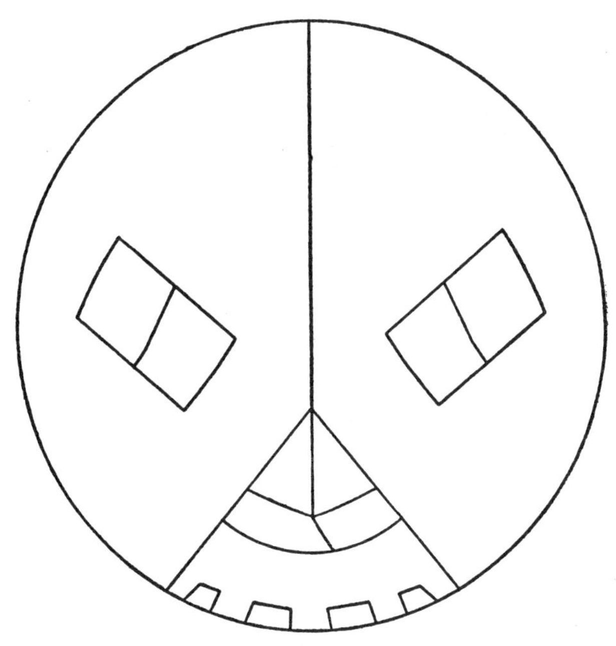

Fig. 1—Jack o' Lantern

FIG. 2—Jack o' Lantern

96. A REMARKABLE OCCURRENCE. The two accompanying diagrams show how the board can be cut into only 11 pieces to form the tableau.

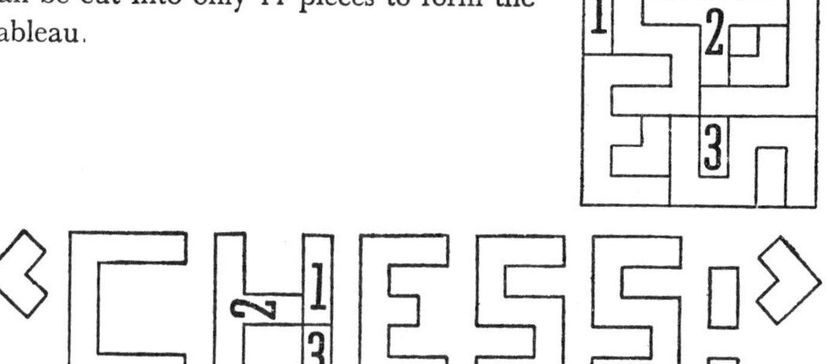

97. TANGRAM PARADOXES. The diagrams given below show the construction of the two men in Fig. 3 (page 50). The man with the foot is actually thinner than his companion without a foot, by the area marked out by the broken line.

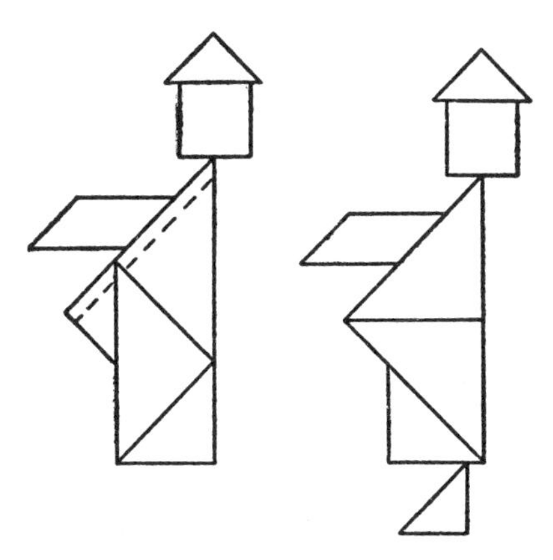

The Man Who Lost a Foot

The next diagram shows how to construct the notorious fish (page 51). And page 172 shows the solution of the paradoxical diamonds (page 52).

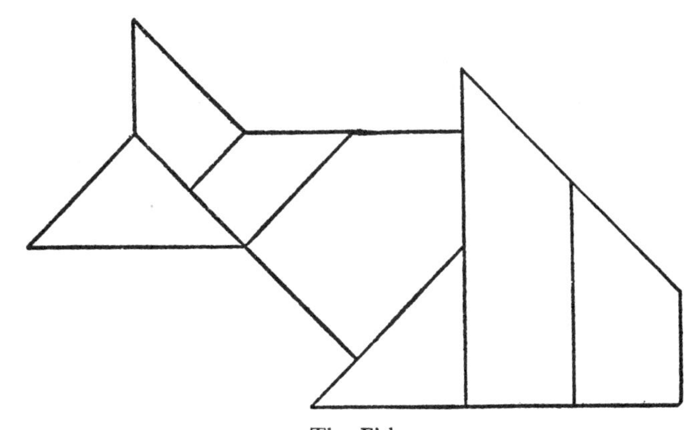

The Fish

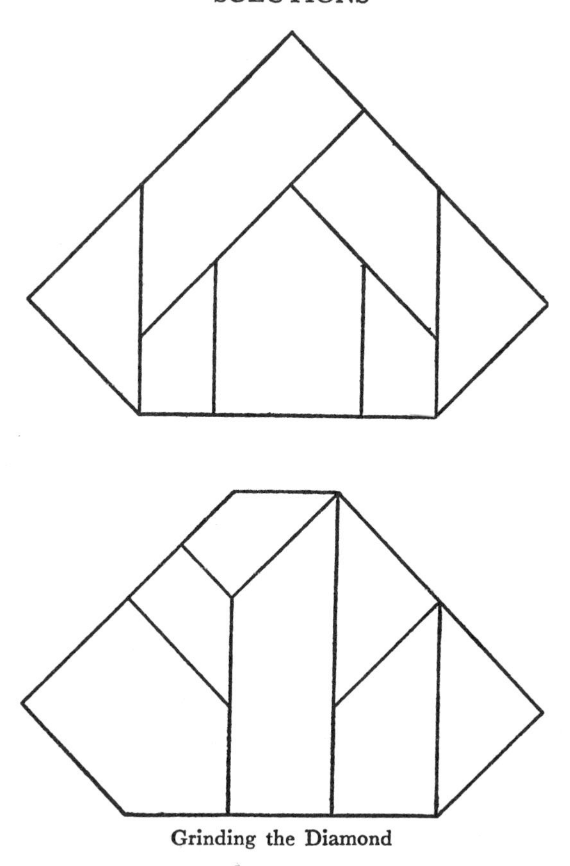

Grinding the Diamond

98. THE CLUB INSIGNIA. If the larger circle is ⅔ of an inch in diameter, the side of the triangle will be $1/\sqrt{6}=0.4082$ of an inch.

99. THE FERRYBOAT GATE. The spread is 14 feet, measured between centers of the outermost rods. The principal diagonal members always make isosceles triangles with legs of 15 inches; when the base is 18 inches, the altitude on this base is 12 inches (remember the 3-4-5 right triangle). Two such triangles per panel for 6 panels (between rods) account for 12 feet of the spread, and 2 feet are added by $12 \times 2''$ distances from center of rod to collar pivot.

100. STRIKING A BALANCE. The lever will balance if the *moments* on each side of the fulcrum are equal. The moment of a force is the product of its magnitude and its distance from the fulcrum. If we call the distance of the 60-pound weight from the fulcrum unity, then the 105-pound weight must be placed $4/7$ of this distance on the other side ($60 \times 1 = 105 \times 4/7$).

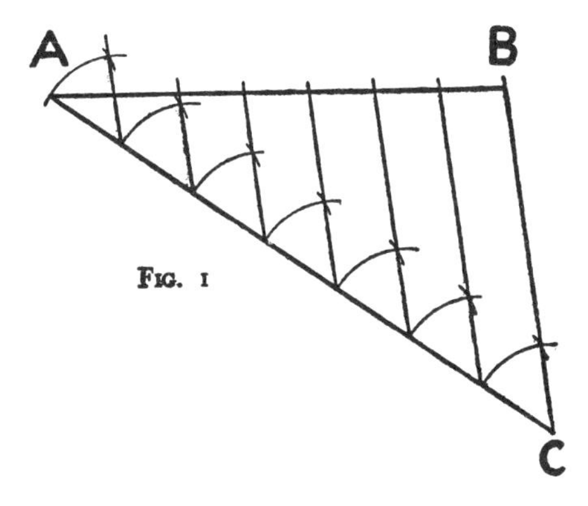

FIG. 1

The method usually given in textbooks for dividing a given line into a given number of segments is illustrated in Fig. 1. To divide line AB into 7 equal segments, draw line AC at any arbitrary angle; lay out 7 equal lengths of any convenient size from A terminating in C; connect CB; through each division point on AC draw a line parallel to CB; these parallels will divide AB into 7 equal segments.

From the practical point of view, the method illustrated in Fig. 2 is much superior. Parallel to the given line AB draw another line CD, at any convenient distance; on CD lay out 7 equal lengths of arbitrary size greater than $1/7$ of AB; connect the terminal points C and D respectively with A and B; these two lines intersect at E; connect E with each of the division points on CD;

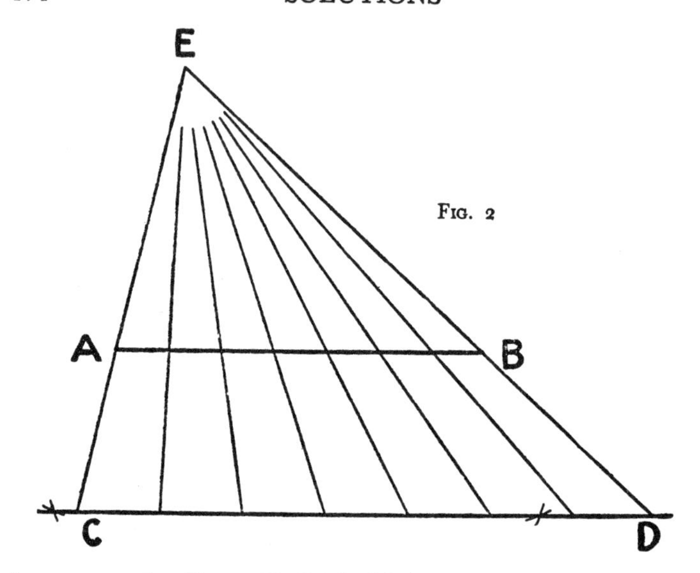

Fig. 2

these connecting lines will divide AB into 7 equal segments.

The second method requires the construction of only one parallel instead of $n-1$ parallels. It is far superior where the construction, for lack of bow compasses, can only be approximated (as is often the case in military sketching). The drawing "by eye" of $n-1$ parallels allows wide error. But where only one parallel need be drawn, at an arbitrary distance, it can be fixed with tolerable accuracy by the opposite edges of a ruler or any suchlike tool.

101. AN INTERCEPT PROBLEM. The length AB is equal to $8r/5$. The computation involves analytical algebra. Determine the equation of the tangent OD and the equation of the middle circle; solve these equations simultaneously to find the points of intersection A, B; then compute the distance between these points.

Take O as the center of rectangular co-ordinates, with OE as the x-axis. The equation of line OD is then given by the ratio of DG (perpendicular from D to OE) to OG. Since ODF is a right triangle, the altitude DG is equal to (OD) (DF) /OF. DF equals r

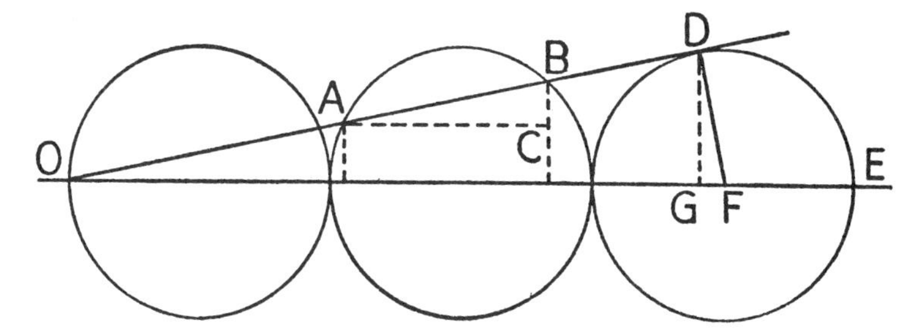

and OF equals $5r$; from these values the other terms can be computed. The equation of OD is found to be

$$\frac{y}{x}=\frac{\sqrt{6}}{12}$$

The equation of the middle circle is

$$(x-3r)^2+y^2=r^2$$

Solve the two equations simultaneously for the value of x, which is found to be

$$x=\frac{72r}{25}\pm\frac{8r}{25}\sqrt{6}$$

The two values of x are the abscissas of the points B and A. The difference between these values, equal to $\frac{16r}{25}\sqrt{6}$ is the length of AC. Since ABC is a right triangle by construction, AB can be computed from AC and BC. The latter can be computed by use of the equation for OD. The desired length AB is found to be $8r/5$.

102. THE BAY WINDOW. The ladder reaches a point 16 feet above the ground. The algebraic solution involves a fourth-degree equation, which is easiest solved by trying out integral values within a narrow range, since the answer obviously is more than 12 feet and less than 20. But easier than solving the equation is to recognize that the problem is based on our old friend, the 3-4-5 right triangle.

103. THE EXTENSION LADDER. In the diagram, DA represents the side of the Mayfield Building, AE the street (at sidewalk level), and CE the face of the building on the opposite side. The extension ladder is represented by DE (the proportions being distorted for the sake of clarity). AC is the traffic barrier ladder. The bracing ladder is not shown, as it contributes nothing to the solution.

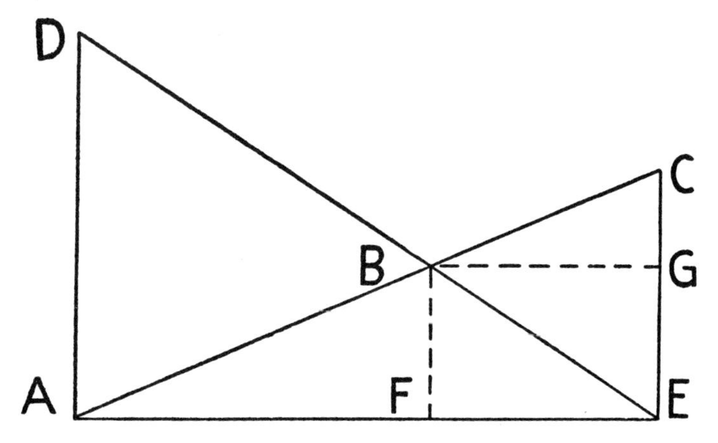

We are told that DE=70 feet less 2 feet 2 inches=67 feet 10 inches=814 inches; that AC=22 feet 11 inches=275 inches; and that distance BF=5 feet 10 inches=70 inches. DA and CE are parallel, both perpendicular to AE. The problem is to determine the length DA.

Obviously, many equations can be derived from the figure, but the manipulation of these equations has proved puzzling to many solvers.

The difficulty arises fundamentally from the fact that the final formula for the length DA is bound to be a quartic—that is, to involve the fourth power of the unknown x. The solution of a quartic equation is certainly not within the scope of elementary algebra, but in this case the necessity can be circumvented by methods discussed in the solution to No. 142—*The Battle of Hastings.*

The solver is likely to make added difficulty for himself by working from more equations than he needs, not all of them independent. The trick is to select just the right facts, so as to avoid getting into a morass of implicit equations.

Since ADE and ACE are right triangles, we can write

$$DE^2 = DA^2 + AE^2 \tag{1}$$

$$AC^2 = CE^2 + AE^2 \tag{2}$$

Subtract (2) from (1).

$$DE^2 - AC^2 = DA^2 - CE^2 \tag{3}$$

Triangles ABD and CBE are similar, whence we derive the proportion

$$DA:CE = AB:BC \tag{4}$$

Now through point B draw line BG parallel to AE. Then triangles ABF and BCG are similar, and we have the proportion

$$AB:BC = BF:CG \tag{5}$$

Combining (4) and (5) gives

$DA:CE = BF:CG$, or

$$DA = \frac{CE \times BF}{CG} \tag{6}$$

Substitute this value of DA in (3).

$$DE^2 - AC^2 = \left[\frac{CE \times BF}{CG}\right]^2 - CE^2 \tag{7}$$

For CE substitute the equivalent expression CG+BF.

$$DE^2 - AC^2 = \left[\frac{(CG+BF) \times BF}{CG}\right]^2 - (CG+BF)^2 \tag{8}$$

Of the four terms in (8), the values of three are given by the statement of the problem, wherefore (8) serves to solve it. Substitute these values, and replace CG by the simpler term x.

$$814^2 - 275^2 = \left[\frac{70^2}{x} + 70\right]^2 - (x+70)^2 \tag{9}$$

The value of the left member of (9) is 662,596—75,625 =586,971. Here we have an integer, the difference between two squares. The squares are not necessarily integral, but it is worthwhile trying out the hypothesis that they are. The method of finding all integral squares that differ by a given integer is expounded in the solution to No. 142—*The Battle of Hastings.*

According to that method, we must find an appropriate term r among the factors of 586,971. The prime factors of this integer are

$$586{,}971 = 3\times3\times7\times7\times11\times11\times11$$

A pair of squares can be found for every factor r that is less than the square root of 586,971, e.g., 1, 3, 7, 9, etc. Since there are many such factors, let us see if we can narrow the search.

In equation (9), x represents the vertical distance from the junction of the two ladders to the raised end of the barrier ladder. Since the latter ladder is nearly horizontal, x is relatively small. Then the term $\left[\frac{70^2}{x}+70\right]$ must greatly exceed the term $(x+70)$. Consequently, r must be relatively large.

For the largest possible values of r we find the following pairs of squares:

FACTORS		r	a	SQUARES
$3\times7\times11$	=	231	2311	1386^2-1155^2
$7\times7\times11$	=	539	551	814^2-275^2
$3\times3\times7\times11$	=	693	155	770^2-77^2

The last pair is the only plausible set. To try it out, we write

$$\left[\frac{70^2}{x}+70\right]=770 \text{ and } (x+70)=77$$

Since the two equations agree in giving the value $x=7$, this is the correct solution of the quartic. By reference to equation (3) it is seen that the first term above gives the value of DA.

The height of the Mayfield Building is thus determined to be 770 inches, or 64 feet 2 inches.

104. THE SPIDER AND THE FLY. Consider the room to be composed of 6 plane rectangles, hinged together in any suitable way so that they can be folded up in the form of a box. If they are unfolded and laid flat in one plane, and the positions of the spider and fly are plotted, then the shortest course will be the straight line joining the two points.

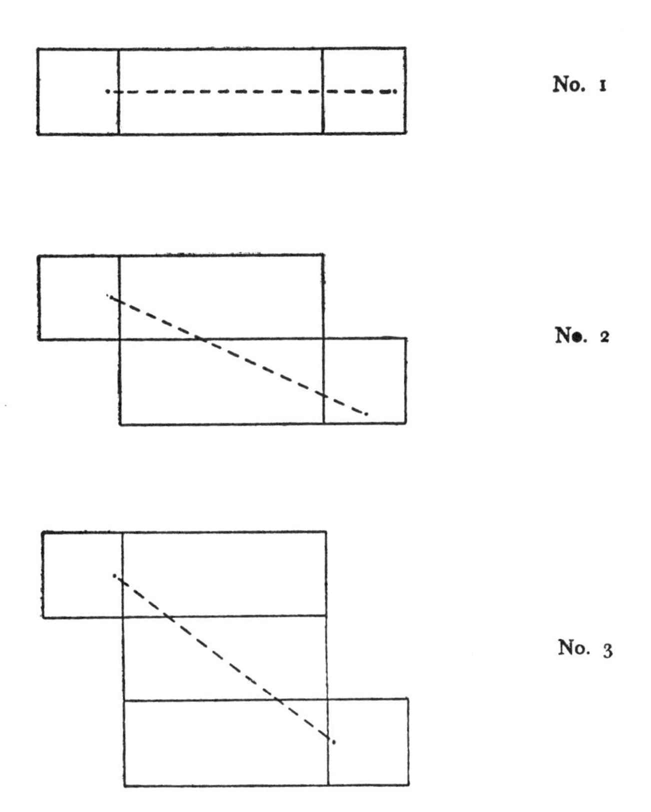

Different ways of hinging the rectangles are possible, and it is a matter of experiment to find which produces the shortest course. Three possibilities are shown in the diagrams. No. 1 requires the spider to travel 42 feet. No. 2 is better, reducing the distance to a little over 40 feet. But the best is No. 3, with a course of exactly 40 feet. The odd fact is that this course touches 5 of the 6 rectangles.

105. THE SPIDER'S COUSIN. The plan of the shortest route is fairly obvious. Since the spider must cross on the floor to reach the inner wall, he cannot do better than head directly for the vertex M in the diagram, at the point where it touches the floor. With one outer wall "opened out" on a hinge with the floor, the first leg of the journey is shown by the line SM. The second leg is shown on the sketch below the pentagon. Two of the inner walls are opened out on a hinge between them, and the shortest route is the diagonal MF.

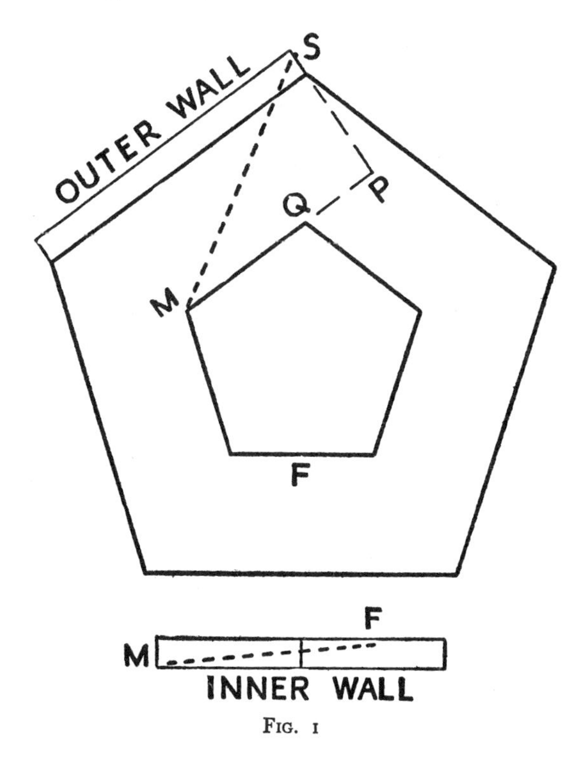

FIG. 1

Computation of the second leg is simple. MF is the hypotenuse of a right triangle whose sides are 9 feet and 1050 feet, the latter being 3/2(700). The height at F being relatively so small, MF can be assumed to be 1050 feet within the "nearest foot" limit. (Actually it is about 1050.04.)

Computation of the first leg is not so simple. SM is the hypotenuse of a right triangle whose sides are MP and PS. The distance PQ is half the difference between the lengths of an outer and an inner wall. Hence MP is 1100 feet. PS is 9 feet plus the distance between the walls; calculation of this distance is the nugget of the puzzle.

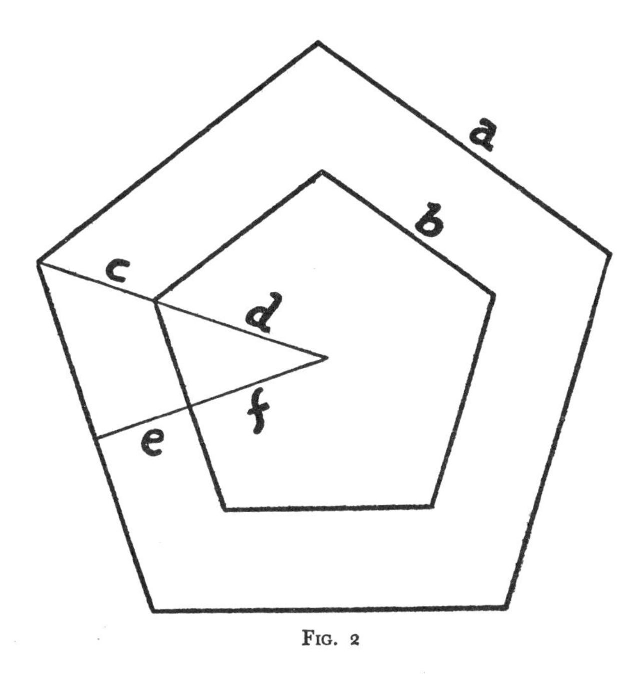

FIG. 2

In Fig. 2, *a* and *b* are the sides of two concentric pentagons and O is the center. The radius of the smaller pentagon is *d;* of the larger, $d+c$. The apothegms (perpendiculars from center to side) are respectively *f* and $f+e$.

The ratio b/d is constant for all pentagons; call this ratio *k*. Now let us find the value of all the segments in terms of *a, b, k.*

We have

$$d=\frac{b}{k} \tag{1}$$

$$c+d=\frac{a}{k} \tag{2}$$

and by combining (1) and (2)

$$c=\frac{a-b}{k} \tag{3}$$

In the small right triangle

$$f^2=d^2-\left(\frac{b}{2}\right)^2=\frac{b^2}{k^2}-\frac{b^2}{4}$$

$$\therefore f=\frac{b}{2}\sqrt{\frac{4}{k^2}-1} \tag{4}$$

By similar triangles we have $e:f=c:d$. Combining this equation with (1), (3), (4) gives

$$e=\frac{a-b}{2}\sqrt{\frac{4}{k^2}-1} \tag{5}$$

In No. 92—*Construction of a Pentagon* we find k to be 1.17558. Then the value of the radical in (5) is 1.3763, and the distance between the walls of the Pentagon Building is 400 times this number, or about 550.52 feet. The distance PS in Fig. 1 is then 559.52 feet. Solving the right triangle gives SM, the first leg of the spider's journey, as almost exactly 1234 feet.

The total length of the route is 1050+1234=2284 feet, correct within a small fraction of a foot.

106. TOURING THE PENTAGON. Imagine a schematic plan in which each corridor is represented by a single line down its center. Each intersection of two or more lines is a *node*. Count the number of *rays* at every node. A ray is a line segment construed to terminate at the node; hence a straight line passing through and beyond the node counts as two rays.

When any *even node* (even number of rays) is reached by a new path, there will always be a new path of retreat from it. But when an *odd node* is entered for the last time, it will necessarily be along the last unused path, and exit can then be made only along a path previously traversed. The number of retracings necessary in this (or any such) puzzle depends solely on the number of odd nodes.

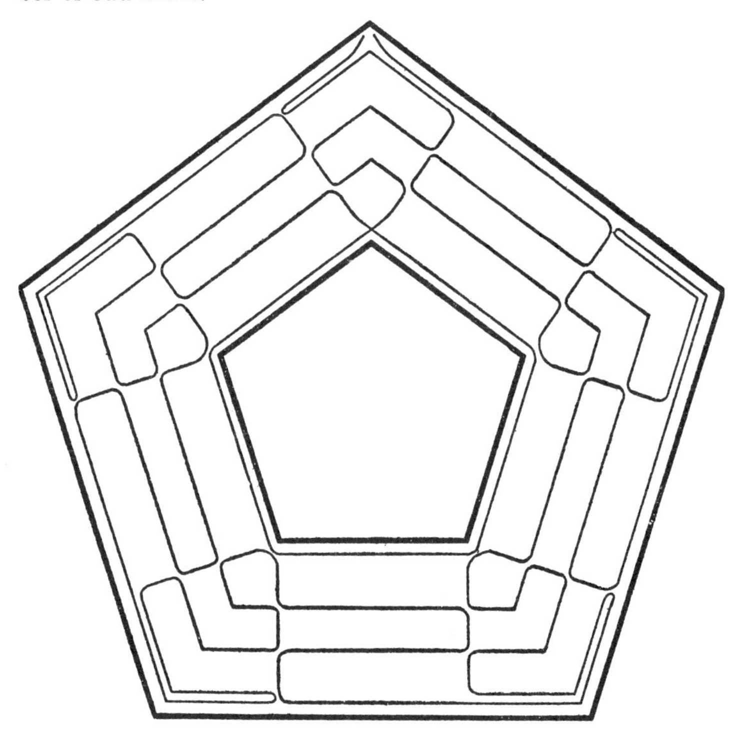

The Pentagon plan has 10 odd nodes, all in the outermost circumferential corridor where it meets the transverse corridors. We shall have to retrace half this number of legs—5. The diagram shows the minimal tour (the continuous line is the course of travel, the partitions being suppressed for clarity). There are many different ways of arranging the circuit of even nodes, but there is no way of shortening the 5 walks "around the corner and back" in the outermost corridor.

107. HOW TO DRAW AN ELLIPSE. It is easy to see that the major axis is just as long as the string, 10 inches. When the pencil point is put on the end of the minor axis, it is at the midpoint of the string. Half of the string forms the hypotenuse of a right triangle, of which the legs are the semi-minor axis and the distance from the center to the focus. The latter distance is half that between the foci, or 3 inches. Half the string is 5 inches. Thus we have the familiar 3-4-5 right triangle. The semi-minor axis is 4 inches, and the whole 8 inches.

108. ROADS TO SEDAN. The given line segments are AB and CD. The problem is to draw a line through point X that would pass through the intersection of AB and CD if all three lines were sufficiently extended.

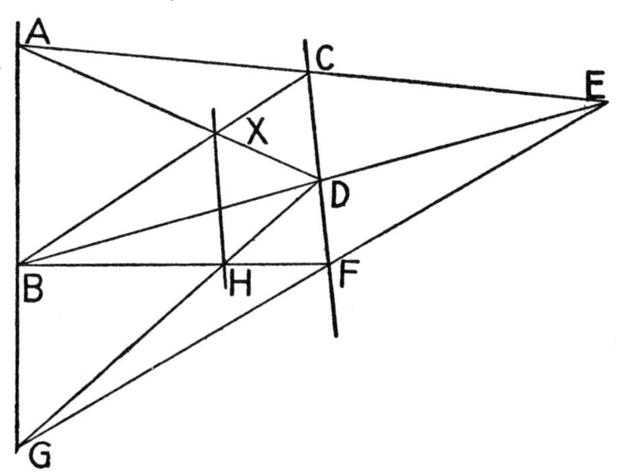

Through X draw any two lines intersecting both AB and CD. The intersections are A, B, C, D. Through AC and BD draw two lines which intersect in E. Through E draw any line intersecting the two given lines in F and G. Connect BF and DG, these two lines intersecting in H. Then XH is a third line of the pencil which intersects at the same point as AB and CD.

109. THE BILLIARD SHOT. Few of the manuals on billiards attempt to explain the calculation of angles, and none employs the rigorous method given here. While the angles have to be learned largely from experience, being much affected by spin and friction, study of the geometry of the ideal frictionless table should be the starting point.

Imagine the table (the rectangle enclosed by the cushions) to be extended in every direction by an endless series of mirror reflections, as illustrated in the diagram on page 186. Imagine the two object balls to be correctly placed in each reflection.

All possible bank shots can be determined by drawing a straight line from the actual cue ball to the reflected object balls, to the point where a simple carom could be made.

The diagram shows 15 of the possible reflected tables. All others more remote are irrelevant; in most of them the object balls are screened by the balls on a nearer table, as the balls in Table 3 are screened by the balls in Table 6.

The line so drawn is the actual direction in which the cue ball must be sent on the actual table. The number of rectangle-lines crossed by the line of aim shows the number of cushions that the cue ball will touch before striking the object balls. Thus, the shots into Table 9 and Table 13 are one-cushion shots; those into Tables 2, 7 and 12 are two-cushion shots.

Of the four three-cushion shots, we will reject that into Table 1 as needlessly long and that into Table 8 as too risky because the carom is so "thin." The shots into Tables 4 and 15 give excellent angles of approach for the carom; our final selection is 4 because this shot is shorter. The complete course of the cue ball can be plotted on the actual table by placing Tables 9, 5 and 4 in turn upon it in proper orientation and transferring the line segments onto the actual table.

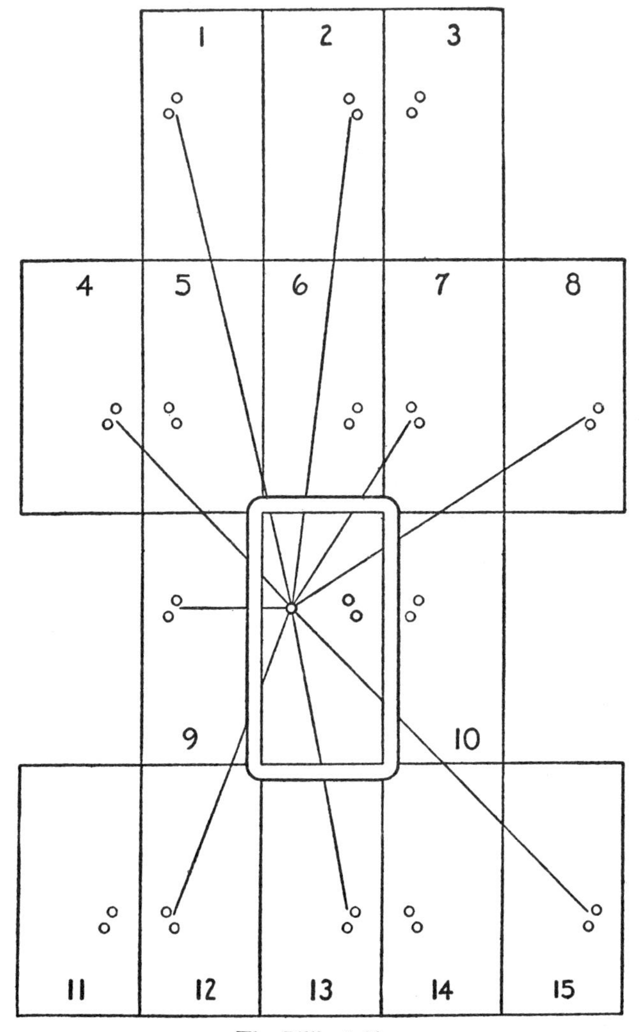

The Billiard Shot

112. THE MISSING DIGIT. The digital root of 673,106 is 5; that of 4,783,205,468 is 2. Since $5 \times 2 = 10$, whose root is 1, the root of the product must be 1. The given digits of the product sum to 69, whose root is 6. To make the root 1, we will have to add 4. Hence the missing digit is 4.

113. FIND THE SQUARE. The units digit of a square number can only be 1, 4, 5, 6, 9, or 0. This fact rules out the 2nd and 5th of the given integers. The digital root of a square is always 1, 4, 7, or 9. The roots of the 1st, 3rd, and 4th numbers are respectively 8, 5, and 3, so that they also are ruled out. Only the 6th, with root of 7, can be a square. (No doubt the reader will like to verify that it actually is a square.)

115. SPELLING ADDITION. Of six solutions, the only one which satisfies the condition that WOES×5=SORER is

```
   9 4 0
   7 3 9
 8 9 2 5
---------
1 0 6 0 4
```

116. ADAM AND EVE. The solution with the maximum total is

```
8384
 803
 626
  50
   8
----
9871
```

The values for D and E (3 and 6) are interchangeable, also for V and O (2 and 5). I find 8 different solutions, or 10 if we include a couple in which A is allowed to have the value zero.

117. RESTORING THE FIGURES. The multiplicand is 74,369,053; the multiplier is 87,956.

118. LETTER DIVISION. The solution is given by substituting each digit for the letter below it:

2	9	7	1	3	5	4	8	0	6
A	B	C	D	E	F	G	H	J	K

119. CRYPTIC DIVISION.

```
99 ) 19107 ( 193
     99
     ---
     920
     891
     ---
      297
      297
      ---
```

120. CRYPTIC MULTIPLICATION.

```
    1475
     677
    ----
   10325
  10325
  8850
  ------
  998575
```

121. CRYPTIC SQUARE ROOT.

```
     3  1  9  4
 √ 10 20 16 36
    9
    -
    1 20
      61
    ----
      59 16
      56 61
      -----
       2 55 36
       2 55 36
       -------
```

122. A COMPLETE GHOST.

First division: 333) 100007892 (300324

Second division: 29) 300324 (10356

123. A GHOST ADDITION. The key to this problem is the fact that every triangular number has a digital root of 1, 3, 6 or 9. Since it is stipulated that the sum must be a triangle not divisible by 3, its root must be 1.

Make a tabulation of the ten possible sequences of five digits and compute the digital root of the sum of each sequence.

SEQUENCE	ROOT
1, 2, 3, 4, 5	6
2 . . .	2
3 . . .	7
4 . . .	3
5 . . .	8
6 . . .	3
7 . . .	7
8 . . .	2
9 . . .	6
0 . . .	1

We have got to select two sequences whose roots sum to a number whose root is 1. We can at once exclude the sequences commencing with 1, 9, 0, for lack of complementary sequences. Evidently the two sequences chosen must have roots 2 and 8 or 3 and 7.

SEQUENCE	ROOT	SUM
2, 3, 4, 5, 6	2	20
3 . . .	7	25
4 . . .	3	30
5 . . .	8	36
6 . . .	3	30
7 . . .	7	25
8 . . .	2	20

To narrow the search further, write down the actual sums of the seven remaining sequences.

Since the sums corresponding to like roots are the same, there are actually only a few combinations to try. If we choose sequences whose roots are 2 and 8, the final total will be 200+36 or 360+20: neither gives a triangle (refer to list of triangular numbers in the Appendix). If we choose roots 3 and 7, the total will be 250+30 or 300+25. Only the latter gives a triangle, 325.

Since either of two sequences whose roots are 3 can be paired with either of two whose roots are 7, there are in all four solutions, as follows:

47	67	43	63
58	78	54	74
69	89	65	85
70	90	76	96
81	01	87	07
325	325	325	325

125. TWO-DIGIT NUMBERS.

(a) 36. (b) 27. (c) 25 or 36. (d) 54. (e) 29, 38, 47, 56.

126. THREE-DIGIT NUMBERS.

(a) 198. (b) None. (c) 189. (d) 629. (e) 111. (f) 132, 264, 396.

127. PRIME NUMBERS. The next three primes are 1009, 1013, 1019. The simplest way to find them is to write out the integers from 998 up, say to about 1020, and strike out those divisible by the successive primes 2, 3, 5, 7, etc. The even numbers may be omitted in the first place. Then those whose digital roots are 3, 6, or 9 are struck out, also those that end with 5. Test for 11, then try actual division by 7, 13, 17, 19, 23, 29, and 31. A simple procedure is to divide each of these primes into 1,000; if there is a remainder, add it to 1,000 to give an integer divisible by the prime. Count forwards and backwards from this integer to strike out the others also divisible by the same prime. After the work has been carried through to 31, the three integers named above are the only survivors.

It is not necessary to go higher than 31, because the square of the next prime, 37, exceeds 1020. If any number on the list were divisible by a prime higher than 31, then it would have another factor lower than 31—and all lower primes have been tested.

128. THE SALE ON SHIRTS. The problem is to factor 60377. The statement hints that one factor is less than 200. The terminal 7 can result only from factors whose terminals are 7 and 1 or 9 and 3. Hence the factors can be found by making trial divisions with primes below 200 that end in 1 or 3 (or 7 or 9). The factors prove to be 173 and 349, both prime. Consequently, there must have been 349 shirts at $1.73 each.

129. A POWER PROBLEM. The 5th root of 844,596,301 is 61. This problem can be solved in short order from the following considerations:

(a) There are nine digits in the given integer; its 5th root consequently is greater than 10 and less than 100.

(b) The terminal digit of the given integer is 1; its 5th root must likewise end in 1.

(c) The digital root of the given integer is 4; the digital root of its 5th root must be 7 (as may be discovered by writing out the roots of the powers of digital roots 1 to 9 inclusive).

130. THE ODD FELLOWS PARADE. The number of Odd Fellows in the parade was 367, the only integer under 497 that satisfies the given conditions.

The following is a general method for attacking all problems of this sort, where an integer is defined by its remainders on divisions by various primes. The method is developed from elementary algebra. Special conditions may lead to short cuts, so that the method does not necessarily have to be followed in full.

The problem states that s, the total of Odd Fellows, is of form

$$s=3a+1=5b+2=7c+3=11d+4 \qquad (1)$$

where a, b, c, d are unknown integers.

From (1), find b in terms of a.

$$b=\frac{3a-1}{5} \tag{2}$$

Since b is integral, the numerator of the fraction must be exactly divisible by 5. By trial we find that the lowest value of a that serves is 2. Consequently a must be of form

$$a=2+5x \tag{3}$$

where x is an unknown integer. From (2) and (3)

$$b=1+3x \tag{4}$$

Any value of x will give an integer that satisfies the conditions as to a and b. But we also have c and d to reckon with. Combine (1) and (3) to find c in terms of x:

$$c=\frac{4+15x}{7} \tag{5}$$

By trial we find that the lowest value of x that makes c integral is 3. Consequently x is of form

$$x=3+7y \tag{6}$$

where y is an unknown integer. Replacing x by its value in terms of y, we derive from (3) (4) (5) three new equations:

$$\begin{aligned} a&=17+35y \\ b&=10+21y \\ c&=\ \ 7+15y \end{aligned} \tag{7}$$

Finally, we have to reckon with d. From (1) and (7) we find that

$$d=\frac{48+105y}{11} \tag{8}$$

from which it follows that y is of form

$$y=3+11z \tag{9}$$

where z is an unknown integer. By substituting its value in terms of y in (7) and (8) we derive a final set of equations

$$\begin{aligned} a&=122+385z \\ b&=\ \ 73+231z \\ c&=\ \ 52+165z \\ d&=\ \ 33+105z \end{aligned} \tag{10}$$

Any value of z will give a set of values for a, b, c, d that will satisfy the problem of the Odd Fellows, but since the answer is limited to $s<497$, we must set z equal to zero and solve for s in (1).

131. THE UNITY CLUB. A short cut is available to solve this problem. The division of the Unity Club marchers by 3, 5, 7, 11 in each case left the array 1 short of having even ranks. The least number in the group was consequently the product $3\times5\times7\times11$, less 1, or 1,154.

132. CINDERELLA TOASTERS. The task is to find the G.C.D. (greatest common divisor) of 389,393 and 831,119. It is not necessary to factor the two integers; G.C.D. can be found "by force." Divide 389,393 into 831,119 to give quotient 2 and remainder 52,333. Divide the remainder into 389,393 to give quotient 7 and remainder 23,062. Continue in the same way, dividing the remainder each time into the previous divisor. The first divisor that gives zero remainder is the required G.C.D. By this method we find the G.C.D. of the given integers to be 887. Since this integer is prime, the price of the toaster must be $8.87, and the number sold at the Main Street Branch is $389393/887 = 439$.

133. SQUADS AND COMPANIES. The Numerian army comprises at least 242,879 men. This number is the L.C.M. (least common multiple) of 1,547 and 34,697. The way to find it is to determine the G.C.D. and multiply the latter by the other factors contained in the two integers.

134. THE MISREAD CHECK. The check was for $51.24.

Call x the dollars and y the cents. Then the amount of the check (in cents) was $100x+y$. The teller first gave $100y+x$, which was $1.11 short of half the correct amount. Hence

$$100x+y=2(100y+x+111) \qquad (1)$$

$$98x=199y+222 \qquad (2)$$

Equation (2) is a Diophantine—it would have an infinity of solutions but for the fact that x and y must be integers. A further limitation is that both integers must be less than 100.

The following method used to solve equation (2) can be applied to many linear Diophantines.

A factor common to all terms of an equation but one must be a factor of that one as well. In (2) the first and last terms are even: hence y also must be even. Replace y by $2z$.

$$49x=199z+111 \qquad (3)$$

This reduction of the equation to lowest terms is not actually necessary, but with more complex examples it often proves a welcome simplification.

Divide (3) by 49, the coefficient of x.

$$x=z\left(4+\frac{3}{49}\right)+\left(2+\frac{13}{49}\right) \qquad (4)$$

In order to make x integral, we must find a value for z such that $3z+13$ is a multiple of 49. The lowest value that satisfies is $z=12$, whence $x=51$, $y=24$.

One solution, then, is \$51.24. Is any other solution possible? Examine (3). When $z=1$, x is about 6. This means that x must be about 6 times z. Now the maximum possible value of x is 99, and of z, consequently, $\frac{99}{6}$ or about 16. From equation (4) we see that the next-higher value of z, beyond 12, that satisfies the equation, is far beyond 16. Hence the solution \$51.24 is unique.

135. TRANSFERRING DIGITS. The required number is enormous, but it can be found by "brute force."

Since we do not know how many digits there are in the required integer, we will represent them by A, B, C... as read *from right to left*. Then the integer is of form

$$A+10B+100C\ldots\ldots+10^{n-1}Z \qquad (1)$$

where n is the number of digits.

Let us take A as the terminal digit to be transferred. When it is put at the other end, the integer becomes

$$B+10C+100D \ldots +10^{n-2}Z+10^{n-1}A \quad (2)$$

The stipulation is that (1) is to be $\frac{4}{5}$ of (2). (Remember that the digits are represented in reverse of the way they are written.) Then

$$A+10B+100C \ldots +10^{n-1}Z = \tfrac{4}{5}(B+10C \ldots +10^{n-2}Z+10^{n-1}A) \quad (3)$$

Clearing of fractions and expanding, we have

$$5A+50B+500C \ldots +5(10^{n-1}Z) = 4B+40C \ldots +4(10^{n-2}Z)+4(10^{n-1}A) \quad (4)$$

Now collect the A terms on the right, all other terms on the left:

$$46(B+10C \ldots +10^{n-2}Z) = A[4(10^{n-1})-5] \quad (5)$$

From (5) it follows that the right-hand member is divisible by 46. In other words, we must find values for A and n such that

$$\frac{A[4(10^{n-1})-5]}{2\times 23} \quad (6)$$

will be integral. Since the expression in brackets is odd, it is not divisible by 2; therefore A is divisible by 2, and we can write

$$A=2, 4, 6 \text{ or } 8 \quad (7)$$

Since A is not divisible by 23, the expression in brackets must be. The expansion of this expression for values of $n=1, 2, 3\ldots$ gives 35, 395, 3995, etc. To find the first of these terms divisible by 23, set up a long division in form

```
23 ) 399......95 ( 17..
     23
     ---
     169                    (8)
     164
     ---
       59 etc.
```

Bring down 9 from the dividend each time, until a remainder of 11 is reached, so that the final 5 can be brought down (since

115=23×5). This turns out to be a lengthy matter, but is mere arithmetic. The smallest quotient obtainable is

173, 913, 043, 078, 260, 869, 565 (9)

By taking A=2, we have the smallest integer that satisfies the conditions:

2, 173, 913, 043, 078, 260, 869, 565 (10)

Three other answers can be obtained by setting A equal to 4, 6 and 8. In each case, as is seen from (6), number (9) has to be multiplied by half of A to make up the balance of the integer.

136. FIGURATE NUMBERS. The reader should perceive that by its very derivation any number in the Pascal triangle is the sum of the number above it and all to the left of that number on the same row. The sum of the first 25 terms of the 4th order is therefore the 25th term of the 5th order. This term can be found by the formula; it is 20,475.

137. LITTLE WILBUR AND THE MARBLES. The smallest possible number is 210. It is easily discovered by empirical methods. The integer must be divisible by at least six different triangular numbers (not including unity). It is therefore the L.C.M., or a multiple thereof, of the six triangles. If we compute the L.C.M. of the smallest triangles, we see that we cannot do better than start with the factors 2, 3, 5, for out of them we can get the triangles 3, 6, 10, 15. Since 30 is not itself a triangle, we will choose a multiple of 30. The lowest multiple that is a triangle is 120, but out of this number we cannot make any additional triangles. The next multiple is 210, and this is found to serve, for besides being divisible by 3, 6, 10, 15, it is also divisible by 21 and 105.

138. HOKUM, BUNKUM AND FATUITUM. The "triangular hexahedron" consists of two triangular pyramids set base to base, one having an edge one unit greater than the other. The problem is then to find the four successive triangular pyramids

A, B, C, D, such that the "hexahedrons" A+B, B+C, and C+D total 31,311. Then

$$A+2B+2C+D=31{,}311 \qquad (1)$$

Let n be the edge of the smallest pyramid, A. Then by formula the number of its units is

$$\frac{n(n+1)(n+2)}{6} \qquad (2)$$

The edge of B is $n+1$, of C, $n+2$, etc. Write equation (1) in terms of n, expand and collect terms, thus deriving the equation

$$\frac{6n^3+45n^2+123n+120}{6}=31{,}311 \qquad (3)$$

This equation may be reduced to

$$n\left(n^2+\frac{15n+41}{2}\right)=31{,}291 \qquad (4)$$

Instead of troubling to solve the cubic equation (3), we can find n by trial, since (4) shows it to be a factor of 31,291. The prime factors of this integer are 13, 29, 83. From (4) it is clear that n^3 is somewhat less than 31,291. The approximate size of n can be gauged by comparing 10^3 and 80^3 with 31,291. The first is too small, the second too large. The choice for n thus falls on 29, and this number proves to satisfy (4).

The values for A, B, C, D are then respectively 4495, 4960, 5456, 5984. But the solver need not compute these numbers unless he wishes to check his work. All that is asked are the "powers" of Hokum, Bunkum and Fatuitum, and for $n=29$ these powers are 30, 31, and 32.

139. SQUARE NUMBERS. The squares can be derived by summing each pair of adjacent numbers of the third figurate order. Every square is the sum of two consecutive triangles, as is clear from the diagram. For analytical proof merely expand

$$\frac{n(n+1)}{2}+\frac{(n-1)n}{2}$$

to show that it is equal to n^2.

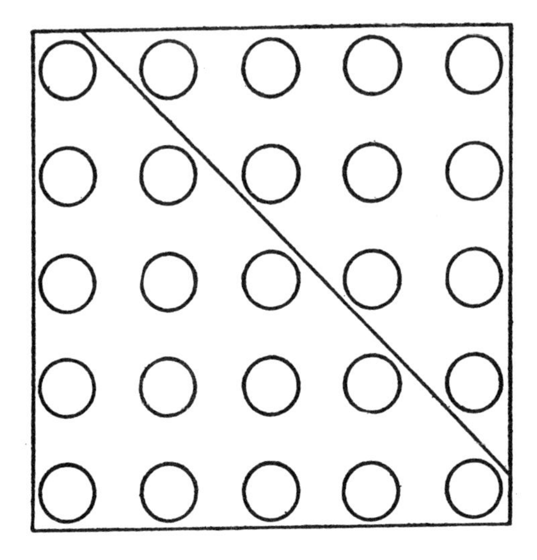

140. SQUARE-TRIANGULAR INTEGERS. One way to solve the problem would be to extend the tables of square and triangular numbers until two more identities are found. But this purely empirical approach can be short-circuited. A little analysis of the nature of the desired integers will show that they can be deduced by a search of the limited tables in this book.

The problem is to find integral values that satisfy the equation

$$\frac{x(x+1)}{2}=y^2 \qquad (1)$$

One of the two integers, x or $x+1$, must be even.

If x is even, it may be replaced by $2n$, and (1) becomes

$$n(2n+1)=y^2 \qquad (2)$$

If $x+1$ is even, it may be replaced by $2n$, and (1) becomes

$$n(2n-1)=y^2 \qquad (3)$$

For convenience we will combine (2) and (3) as

$$n(2n\pm 1)=y^2 \qquad (4)$$

Equation (4) states that y^2 is the product of two factors, each necessarily different from y. It follows that y must comprise at least two factors ab, and that *

$$n=a^2 \qquad 2n\pm1=b^2 \qquad (5)$$

The desired integers can be determined by searching our table of squares for pairs of integers that satisfy (5).

The search can be narrowed by consideration of the terminal digits. Thus:

If the terminal of n is............	0	1	4	5	6	9
then the terminal of $2n+1$ is......	1	3	9	1	3	9
then the terminal of $2n-1$ is......	9	1	7	9	1	7

We can eliminate the cases that result in 3 and 7, since neither digit can be the terminal of a square. Then we need only examine squares ending in 0, 4, 5, 9 $(=n)$ to see if $2n+1$ is a square, and squares ending in 0, 1, 5, 6 $(=n)$ to see if $2n-1$ is a square.

We need go no higher than 29^2 to find four different squares that satisfy the conditions for n in (5).

n	$2n+1$	$2n-1$	y^2 TRIANGLE	SQUARE
4	9		$36=8.9/2$	$=6^2$
25		49	$1225=49.50/2$	$=35^2$
144	289		$41{,}616=288.289/2$	$=204^2$
841		1681	$1{,}403{,}721=1681.1682/2$	$=1189^2$

*The reader may ask why a different distribution of the factors is not possible, e.g.

$$n=ac^2 \qquad 2n\pm1=ab^2$$

It is easily proved that n and $2n\pm1$ cannot contain a common factor, The above equations can be written as

$$\frac{n}{a}=c^2 \qquad (6)$$

$$\frac{2n}{a}\pm\frac{1}{a}=b^2 \qquad (7)$$

Equation (6) states that a is a divisor of n. Then the first term of (7) is integral. Also, b^2 is integral. But the term $\frac{1}{a}$ cannot be integral, since $a>1$. Hence (6) and (7) cannot be reconciled, and it follows that n and $2n\pm1$ are relatively prime.

141. PARTITION OF A TRIANGLE. The truth of the proposition can be seen intuitively from the diagram. Here two different triangular numbers are represented by triangles of cannon balls. Each can be dissected into three triangles, of which the outer three are equal, while the central triangle is the next-smaller or next-larger according as the side of the large triangle is even or odd.

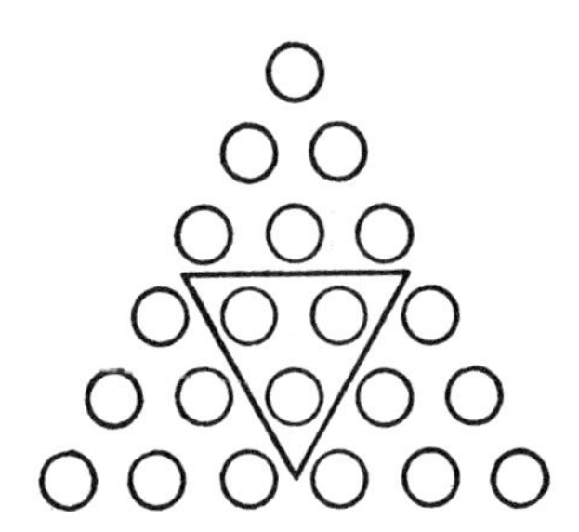

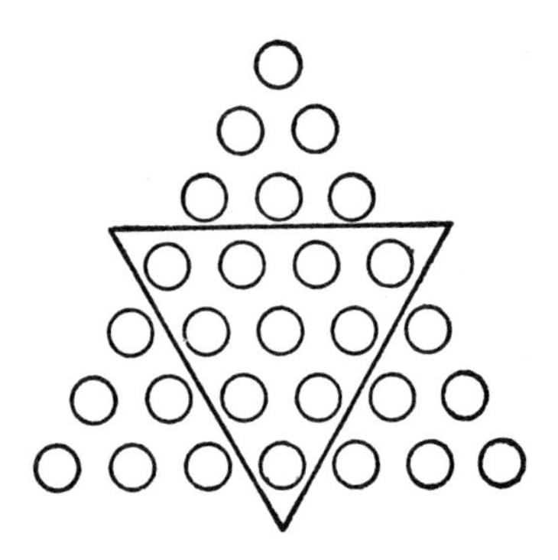

The central triangle plus one of the outer ones equals a square. The large triangle is thus the sum of this square and the two remaining smaller triangles.

For analytical proof, show that the sum of the square of n and two triangles of $n-1$ is $n(2n-1)$, which can be written $\frac{2n(2n-1)}{2}$ and thus is seen to be the triangle of $2n-1$. Similarly show that the sum of the square of n and two triangles of $n+1$ is $n(2n+1)$, which can be written $\frac{2n(2n+1)}{2}$ and so is seen to be the triangle of $2n$. Now, the two expressions $2n-1$ and $2n$ give the complete series of integers as n varies from unity to infinity, so that their triangles comprise all of the triangular numbers. Hence any triangular number p can be partitioned into (a) a square of $p/2$ and two triangles of $(p/2)-1$; or (b) a square of $(p+1)/2$ and two triangles of $(p-1)/2$.

142. THE BATTLE OF HASTINGS. The problem is to find two squares that differ by 512—"half a thousand footmenne and full douzaine more of knights."

The general problem of finding integral values for

$$x^2 - y^2 = C$$

is solved by factoring the left side into $(x + y)(x - y)$, then equating these two terms with pairs of factors of C (a given integer). Each different pair of factors yields a different solution.

The available pairs of factors of 512 ($= 2^9$) are 256 and 2, 128 and 4, 64 and 8, 32 and 16. (The pair 512 and 1 is not available, for the reason that the factors must be both even or both odd, since their sum is $2x$ and difference $2y$, even in both cases.) These factors generate the following solutions:

$$129^2-127^2=16641-16129=512$$
$$66^2-62^2=4356-3844=512$$
$$36^2-28^2=1296-784=512$$
$$24^2-8^2=576-64=512$$

Any one of the sets would answer to the description of the two armies so far as concerns the difference of 512. But we are further told that after slaying half the foe with the loss of "only a few score" of their own men, the Saxons reduced the armies to equality. The only numbers that satisfy this statement are 1296 (Normans) and 784 (Saxons).

143. THE DUTCHMEN'S WIVES. The amount spent by each individual is a square number, and the difference of the expenditures within each family is 63 shillings. The first step is to find three sets of squares that differ by 63. The method is explained in No. 142—*The Battle of Hastings.* The required numbers are

$$32^2-31^2=63$$
$$12^2-9^2=63$$
$$8^2-1^2=63$$

The integers in the first column represent expenditures by the husbands; in the second column, by the wives. Now we have to pick out the integers that differ by 23 and 11. It is easily seen that Anna (31) is the wife of Hendrick (32); Katrun (9) is the wife of Elas (12); Gurtrun (1) is the wife of Cornelius (8).

144. THE CRAZY QUILT. The quilt made by Mrs. Thompson and Effie together was made by combining the pieces of two smaller squares. Subsequently, the large square was dissected into two smaller squares, different in size from the original components. The problem is then to find integral values that will satisfy the equation

$$A^2+B^2=C^2+D^2=E^2 \tag{1}$$

Since there must be an infinity of solutions, we will seek that for which E is the smallest possible integer.

The equations (1) may be written

$$\begin{aligned} E^2-A^2&=B^2 \\ E^2-C^2&=D^2 \end{aligned} \tag{2}$$

Refer to the method used in the solution of No. 142—*The Battle of Hastings.* The first of the equations (2) may be written:

$$(E + A)(E - A) = B^2 \tag{3}$$

The right-hand term must contain two equal factors, besides the equal factors B. Then B itself must be composite, say with factors r and b, and the terms of (2) can be equated:

$$E + A = rb^2, E - A = r \tag{4}$$

Solving for E and A we have:

$$E = \frac{r(b^2 + 1)}{2}, \qquad A = \frac{r(b^2 - 1)}{2} \tag{5}$$

By similar operation on the second of the equations (2), we derive:

$$E = \frac{q(d^2 + 1)}{2} \tag{6}$$

where q and d are the factors of D. Combining the values of E given in (5) and (6) gives:

$$r(b^2 + 1) = q(d^2 + 1) \qquad (7)$$

The problem of finding the minimum solution of equations (1) resolves itself into finding the minimum solution of (7). But in the nature of the problem, no solution of (7) is satisfactory unless it gives four different values to A, B, C, D. This in turn is possible only if the values of r, b, q, d are all different. But then r, for example, must be a factor of the right-hand member $q(d^2+1)$ different from either q or (d^2+1). Consequently this member must contain yet another factor x, and we have two alternatives:

$$\text{Either} \quad q=xr \text{ and thus } x=\frac{b^2+1}{d^2+1} \qquad (8)$$

$$\text{or} \quad d^2+1=xr \text{ and thus } x=\frac{b^2+1}{q} \qquad (9)$$

The rest is a matter of experimenting with the lowest possible integers that will satisfy either (7) or (8). It will be found that the minimum solution is given by the following values in connection with (8):

$$x=10,\ r=5,\ b=3,\ q=1,\ d=7$$

The values generated in equation (1) are A=20, B=15, C=24, D=7, E=25. That is,

$$20^2+15^2=400+225=625$$
$$24^2+\ 7^2=576+\ 49=625$$

The next-higher possible value of E is considerably more than 625, as the reader will no doubt realize from the empirical step in this solution. A quilt composed of 625 3-inch squares measures 6 feet 3 inches on each side (no allowance for overlap). This seems "right for the four-poster bed," and the next-larger size would be absurdly large. We may justly conclude that *The Crazy Quilt* intended is the minimum of 625. The wording makes clear that Effie took away 24^2 of the combined 25^2 quilt, leaving Mrs. Thompson with 7^2, which measured only 21 inches on each side.

145. THE FOUR TRIANGLES PROBLEM. The problem is to find four right triangles such that (a) all four have one side of a given length k; (b) all sides of all triangles can be expressed in integers; (c) the total of the four perimeters is the minimum possible.

If p and q are the legs of a right triangle and k is the hypotenuse, then $p^2+q^2=k^2$. Since the side of the given square k may be used as either hypotenuse or leg of any of the triangles, our task is to find at least four different integral sets of solutions for the equation

$$k^2=p^2\pm q^2 \qquad (1)$$

Let us first try to find the minimal solution where k is always a leg, never a hypotenuse. Then we need deal only with the minus sign in (1).

As is shown in No. 144—*The Crazy Quilt,* to obtain multiple solutions of (1) we must make k composite. If we make it even, we must make it doubly even, that is, divisible by 4. If we choose 4 as the minimal even factor, we find that the least other factor available, to give four different pairs of factors for k^2, is 3.

These pairs are:

$$\begin{aligned} k^2 = 12^2 &= 2 \times 72 \\ &= 4 \times 36 \\ &= 6 \times 24 \\ &= 8 \times 18 \end{aligned}$$

The corresponding triangles are shown in the diagram:

37–35–12; 20–16–12; 15–12–9; 13–12–5.

The number of matches required is the sum of these numbers, 198.

Let us examine all other possibilities to see if this is a minimal solution. Suppose we take k^2 as odd. Then the least factors we can assign to k are 3 and 5. The available pairs of factors of 225 are 1 and 225, 3 and 75, 5 and 45, 9 and 25. These generate the triangles:

113–112–15; 39–36–15; 25–20–15; 17–15–8.

Obviously this solution has a much larger perimeter than the first. We can economize by replacing the largest triangle with 15–12–9, where the side of the square is used as a hypotenuse of the triangle. But even then, the number of matches required is 226.

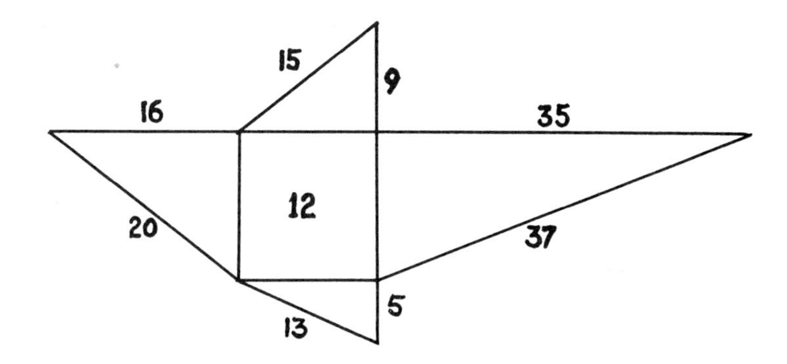

Finally, let us consider the possibility of making k a hypotenuse more than once. Then equation (1) becomes:

$$k^2 - p^2 = q^2 \qquad (3)$$

Then we have:

$$2k = a(b^2 + 1),\ 2p = a(b^2 - 1) \qquad (4)$$

where $q = ab$, a composite number. The least value we can assign to b is 2, making k at least 5. This gives the familiar 5–4–3 triangle. To obtain a second solution for any given k, we will have to make k composite, with a least factor 5.

Obviously any such solution would create a larger perimeter for the figure than the solution $k = 12$.

147. THE FIVE-SUIT DECK. Herewith is given the table of partitions carried up to $n=16$. As applied to the five-suit deck, it shows that there are 97 patterns of bridge hands—the partitions of 16 into from one to five parts inclusive, which total 101, less 4 that cannot occur

16	0	0	0	0
15	1	0	0	0
14	2	0	0	0
14	1	1	0	0

If a small portion of the table is made by writing out the actual partitions and counting them, this fact will be realized:

If we are to partition an integer, say 11, into 3 parts, we must start by putting at least 1 unit in each part: 1 1 1. That leaves 8 units to be distributed in all possible ways. But we have already tabulated the ways of partitioning 8 into 1, 2, or 3 parts. All partitions of 11 into exactly 3 parts are given by this total of ways to break up 8 into not more than 3 parts and then add them into our base 1 1 1.

The correct entry to make in the cell, column 11, row 3, will therefore be found by summing the entries in column 8 down to row 3 inclusive. Or, to generalize, the entry in column n, row r, is found by summing column $n-r$ from top down to row r inclusive.

$n =$	1	2	3	4	5	6	7	8	9	10	11	12	13	14	15	16
1	1	1	1	1	1	1	1	1	1	1	1	1	1	1	1	1
2		1	1	2	2	3	3	4	4	5	5	6	6	7	7	8
3	1		1	1	2	3	4	5	7	8	10	12	14	16	19	21
4		2		1	1	2	3	5	6	9	11	15	18	23	27	34
5			3		1	1	2	3	5	7	10	13	18	23	30	37
6				5		1	1	2	3	5	7	11	14	20	26	35
7					7		1	1	2	3	5	7	11	15	21	28
8						11		1	1	2	3	5	7	11	15	22
9							15		1	1	2	3	5	7	11	15
10								22		1	1	2	3	5	7	11
11									30		1	1	2	3	5	7
12										42		1	1	2	3	5
13											56		1	1	2	3
14												77		1	1	2
15													101		1	1
16														135		1
															176	
																231

To fill column 11 in order from top to bottom, we must sum, in order, portions of columns 10, 9, 8 . . . 2, 1. The terminal cells to be included in each summation lie on a minor diagonal extending downward-left from cell, column 10, row 1. And so for any column *n:* to write the entries from top to bottom we sum successive columns leftward, down as far as the minor diagonal that meets the top edge in column $n-1$.

Of course the minor diagonal sooner or later crosses the principal diagonal, and beyond the intersection we must sum each column entire down to the principal diagonal. After we have written 10 in column 11, row 5, by a summation of most of column 6, we find that the rest of the numbers to be written are identical with those at the bottom of column 10.

It follows that along a major diagonal (parallel to the principal diagonal) there must come a point sooner or later where all subsequent entries are identical. The first of such identical numbers always occurs in an even column, and the row number is half the column number.

The integers at which the major diagonals begin to repeat: 1, 2, 3, 5, 7, 11 . . . are the same as the totals of partitions as n increases from 1 up. The reason for this circumstance is patent from the way in which the table is built.

A point of practical importance in making the table is that it is not actually necessary to sum columns of figures from the minor diagonal up. The correct entry for column n, row r, can be found by summing just two cells: column $n-r$, row r (the cell where the minor diagonal crosses the row) and column $n-1$, row $n-1$ (the cell diagonally adjacent at upper left). For example, in column 16, row 7, the integer 28 is determined as the sum of the adjacent 26 and the 2 in row 7, column 9.

148. ORDERS OF INFINITY. Let the integers in ascending order, 2, 3, 4 . . . be the *indices* of certain sub-groups. Each sub-group is to be composed of fractions, with the numerator and denominator of each fraction summing to the index. For example, under index 5 we write the fractions: $4/1$, $3/2$, $2/3$, $1/4$. Since the fractions are formed by partitioning each index into two parts in all possible ways, and since the number of such partitions of any integer is finite, the number of fractions in each sub-group will be finite.

Within each sub-group the terms can be arranged in any consistent order, e.g., with numerators in descending order of magnitude.

Since the indices 2, 3, 4 . . . go on to infinity, the totality of sub-groups so formed must embrace all the rational numbers, yet the arrangement is discrete.

149. TURKS AND CHRISTIANS. The diagram shows the necessary arrangement. The black circles represent the Christians.

Counting the indicated man as "one," it will be found that the Turks will be counted out first, in the order indicated by the numbers.

The puzzle can of course be solved purely mechanically. Place 30 dots in a circle, mark your starting point, decimate by 13, and note which are the first 15 of the dots to be counted out.

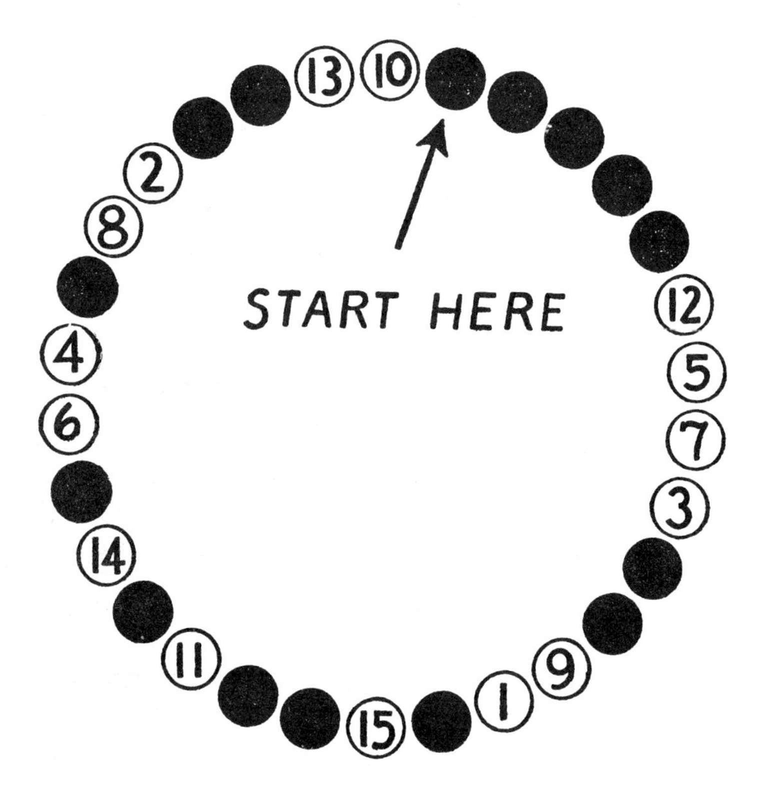

Puzzle makers have expended considerable ingenuity in devising decimation problems that cannot be solved mechanically. The remaining puzzles in this chapter are selected to show some different methods that have been devised.

150. BOYS AND GIRLS. The decimation for the girls must first take out Nos. 1, 2, 3, 7, 10—of course, not necessarily in that order. The decimation for the boys must first take out Nos. 4, 5, 6, 8, 9. The problem can be solved by writing out the order of decimation of a circle of 10 by intervals of 2, 3, 4, etc. Comparatively few figures, however, need be written. As soon as any decimation is found to contain, within its first five numbers, integers from *both* of the above groups, that interval may at once be abandoned as unsuitable. It does not take long to discover that the decimation to take out the boys is 14, for the girls 13.

151. NATIVES AND BRITONS. The two decimations must in one case first count out Nos. 2, 4, 5, 7, 9 and in the other Nos. 1, 3, 6, 8, 10. The problem can be solved by writing out and examining the first five numbers of the decimation of a circle of 10 by 2, 3, 4, etc. But, since we are free to choose any starting point in each case, we shall have to determine for each set of five numbers whether it can be made equivalent to one of the above groups by a cyclical substitution. The easiest way to do this is work the whole solution geometrically. Draw a circle, mark and number 10 points on it, and mark them as natives or Britons by noughts and crosses, in the given pattern. Plot each separate decimation, by 2, 3, 4, etc., on a separate circle, marking simply the first 5 points reached. Compare each such decimation with the original pattern, to see if in any position of rotation the 5 marked points can be made to coincide with the five natives or the five Britons.

It will be found that decimation by 11 gives the series 1, 3, 6, 10, 8 . . . This will serve the chief to count out the Britons first, if he starts counting at No. 1. The decimation by 29 gives 9, 1, 6, 7, 4 which is equivalent to 7, 9, 4, 5, 2. This was the interval the Britons had in mind, with the count commencing on the native No. 9 as "one."

152. JACK AND JILL. The number by which both circles were decimated to leave Jack and Jill to the last was 11. The solution is a matter of writing out the decimations of a circle of 5 and a circle of 4 by 2, 3, 4, etc., until the first interval is encountered that leaves No. 5 and No. 4 respectively to the last.

153. OUT AND UNDER. Like No. 149—*Turks and Christians,* this puzzle can be solved mechanically. Place the cards face up on the table in the desired final order, then pick them up backwards. That is, take up one card, face down. Take up another, place it face down on top of the first, and then transfer the bottom card to the top. Continue in the same way until all the cards have been picked up. The order of the cards, from top to bottom, will then be: A Q 2 8 3 J 4 9 5 K 6 10 7.

The questions asked in No. 154—*The Nightmare* similarly could be answered by working backwards, but I have taken the precaution of making the numbers so large that solving "by hand" is scarcely feasible.

154. THE NIGHTMARE. On the first run through the deck, all odd cards are thrown out, leaving 485 even cards from No. 2 to No. 970 inclusive. On the second deal, since No. 971 was put out, No. 2 will go under and No. 4 will go out. All the rest that go out will be multiples of 4. To put it in tabular form:

	Nos. that go out	*Nos. that go under*
1st deal:	$1+2n$	$2+2n$
2nd deal:	$4+4n$	$2+4n$

The variable n is to take all values 0, 1, 2 . . . up to the highest practicable—beyond which point the term defines numbers higher than any card remaining in the deck. Do not overlook that n must start at zero.

We can build up on this plan a tabulation of what numbers go out and what remain on every deal until the deck is exhausted. Although the deck is in effect a circular arrangement, we have no difficulty in distinguishing what we call the separate "deals." A deal begins whenever the cards are in order of ascending magnitude from the top of the deck down, and ends when they next revert to that order.

I give the complete table for the deck of 971 cards. The manner in which it is built up enables us to generalize; we can make a table for a deck of any number of cards by following these rules of operation:

Deal no.	*Cards in deck*	*Nos. that go out*		*Nos. that go under*	
1	971	1+	$2n$	2+	$2n$
2	485	4+	$4n$	2+	$4n$
3	243	2+	$8n$	6+	$8n$
4	121	14+	$16n$	6+	$16n$
5	61	6+	$32n$	22+	$32n$
6	30	54+	$64n$	22+	$64n$
7	15	86+	$128n$	22+	$128n$
8	8	22+	$256n$	150+	$256n$
9	4	150+	$512n$	406+	$512n$
10	2	406		918	
11	1	918			
	1st	*2nd*	*3rd*	*4th*	*5th*

1st column: At top write total number of cards in the deck. Each other entry is half of the entry immediately above it; in dealing with odd numbers, alternate in writing the smaller and the larger half, writing the *smaller half* on the first occasion an odd number is encountered.

3rd and 5th columns: The coefficients of n are successive powers of 2, commencing with 2^1.

4th column: Each entry either *repeats* that just above it or *changes* to total of coefficients in 4th and 5th columns on row above. It *repeats* (a) when 1st column shows the smaller half of an odd entry above it, and (b) when 1st column shows exact half of an even entry above it, and on said row above, the entry in 2nd column is larger than the entry in 4th column. When converse conditions apply, 4th column *changes.*

2nd column: When 4th column repeats, entry in 2nd column is the sum of the coefficients in 4th and 5th columns on row above; when 4th column changes, entry in 2nd column repeats previous entry in 4th column

From the tabulation we see the answer to the demon's first question: The last card to come out is No 918.

The second question is: When does No. 288 come out? To answer this we must find the algebraic form in the GO OUT columns that define 288. First ascertain how many times 2 is contained in 288; $288=32\times9$. Look for an algebraic form divisible by 32. There is none, which means that the cards that are multiples of 32 do not monopolize a deal. Try 16, the next lower power of 2. Again we find that multiples of 16 do not monopolize a deal, nor do multiples of 8. It is not until we get down to 4 that we find a factor common to 288 and to one of the forms in the GO OUT columns. The second deal takes out numbers of form $4+4n$, and this is the only form that fits 288. If we write $288=4+4n$, then $n=71$. Since the n values start at zero, 71 is the 72nd value it takes, and No. 288 is the 72nd card cast out on the second deal. As 486 cards go out on the first deal, No. 288 goes out as the $486+72$ or 558th card.

The third question is: What is the 643rd card to go out? Since 486 cards go out in the first deal, the 643rd card out will be the $643-486$ or 157th of the second round. For this ordinal number $n=156$. Then $4+4(156)=628$, the number on the 643rd card cast out.

Anyone with a passion for analysis may be interested in deriving formulas that will answer questions of the three types above without the necessity for making a table. Here as a beginning is something he may like to verify. If s is the total of cards in the deck, then the number f on the last card to go out is

$$f=2s-2^n$$

where 2^n is the highest power of 2 less than $2s$. For example, for the deck of 971 cards, $f=1942-1024=918$.

156. A COMMON MISTAKE. Smith and Jones took no account of the ways in which suit-names can be attached to the digits of the pattern to make different combinations. According to their computation, the following two hands are identical, whereas obviously they are different:

♠ A K 7 3 2 ♡ Q 10 9 2 ♢ 6 5 ♣ J 8
♡ A K 7 3 2 ♢ Q 10 9 2 ♣ 6 5 ♠ J 8

Each of their two numbers must be multiplied by a coefficient representing the number of ways in which the names of the four suits can be permuted against the pattern. This number varies according to the number of *identical integers* in the pattern.

For 5 4 2 2, we have a choice of 4 names for the suit of 5; of 3 remaining names for the suit of 4; then there is no further choice. The coefficient here is $4 \times 3 = 12$.

For 5 4 3 1, which has no like digits, the coefficient is $4 \times 3 \times 2 = 24$. The ratio between the two coefficients is 1:2 in favor of pattern 5 4 3 1. Consequently the ratio between the two types of hands is not 36:22 but 36:44 or 11:9 in favor of the 5 4 3 1 pattern.

157. THE ANAGRAM BOX. There are 1,578,528,000 ways. In dealing with the A's, be sure to reckon the permutations of 3 out of 30, not merely the combinations. The puzzle asks "How many ways to pick out *and arrange*," implying that each different permutation of the same three A's among the three positions is to be considered a different arrangement.

158. MISSISSIPPI. The number of permutations is 508,-722,691,276,800.

159. POKER DICE. The table is as follows:

Five of a kind	6
Four of a kind	150
Full house	300
Straight	240
Three of a kind	1200
Two pairs	1800
One pair	3600
No pair	480
$6^5 =$	7776

The reader may have noticed that card probabilities are reckoned from the *combinations* of 5 out of 52 cards, whereas I spoke of 7776 as the number of *permutations* that could turn up on 5 dice. This distinction points the way to what is in my opinion the most accurate way to reckon the dice odds.

We can deal with *combinations* of 52 cards because each card has a separate and unchanging identity. We cannot so deal with 5 dice, because each die can have 6 different identities. Nor can we reckon as though the dice were 6 sets of 6 cards each, numbered in each set 1, 2, 3, 4, 5, 6—a total of 36 cards. To do so would fall into the danger of counting impossible combinations, e.g., an ace on one die combined with a deuce on the same die.

The safe way to count is first reckon the number of *combinations* of five integers out of 1, 2, 3, 4, 5, 6 (allowing repetitions) that satisfy the definition of the hand, then multiply this figure by the number of ways any one combination can be *permuted* on the upper faces of five dice of separate identity.

To illustrate the process, let us take the full house. The triplet may be in any one of 6 denominations, after which the pair may be any one of five. The number of combinations is $6 \times 5 = 30$. Each combination can turn up on the dice in several ways, for example (naming the dice by letters):

A	B	C	D	E
4	4	4	3	3
4	4	3	4	3
4	3	4	4	3

The factor for the permutations of each combination is calculated by the formula $P^n_{a,\,b\ldots} = \dfrac{n!}{a!\ b!\ldots}$ The symbol on the left side means "the number of permutations of *n* objects, of which *a* are of one kind, *b* of another kind, and so on, the sum of $a+b$ . . . being *n*."

In the case of the full house, the factor is given by $\dfrac{5!}{3!\ 2!} = 10$, and the total number of full houses is $30 \times 10 = 300$.

In the case of the straight, there are only two combinations, low straight and high straight; each can be permuted in 5! ways; hence the number of straights is $2 \times 5! = 240$.

It will be noticed that the ranking of hands in poker dice is incorrect, if the basis of ranking be conceded valid. "No pair" should rank just after full house and the straight should rank just ahead.

160. THE MINIM PUZZLE. The total number of paths is 172. This number counts as different the two directions in which the same five cells can be traversed, as is implied by the statement of the problem.

The puzzle is merely an exercise in orderly counting. Reckon the number of paths commencing at one M of each type: Upper left corner, 22; upper right corner, 10; central, 44; intermediate diagonal, 32. There are two M's of each type except the central, so the grand total is given by

$$2(22+10+32)+44=172$$

161. THE SPY. Any given street intersection X can be reached only from the adjacent intersection (if any) to the north or the adjacent intersection (if any) to the west. The number of ways to reach X is the sum of the ways to reach these two adjacent intersections. As this proposition is perfectly general, we can start with the intersection just inside the northwest gate and write the number of routes to each other intersection by a summation process. In doing so the reader will quickly discover that he is compounding "Pascal's triangle," given under No. 136—*Figurate Numbers.* He can simply turn to that tabulation, mark off a rectangle in the upper left which is 10 columns wide and 9 rows deep, and look for 715 in that area. The integer appears only once —the 10th cell of the 5th row. Hence, the spy's office was located in the street adjacent to the east wall, at its intersection with the fifth street down from the north wall.

162. HOW MANY TRIANGLES? The following is not the only way to count the triangles, but it is offered as an example of orderly procedure. Without a rigorous procedure, it is easy to miss some triangles or count some twice over.

The 57 points of intersection are labeled with letters, all points of the same letter standing at the same distance from the center, B.

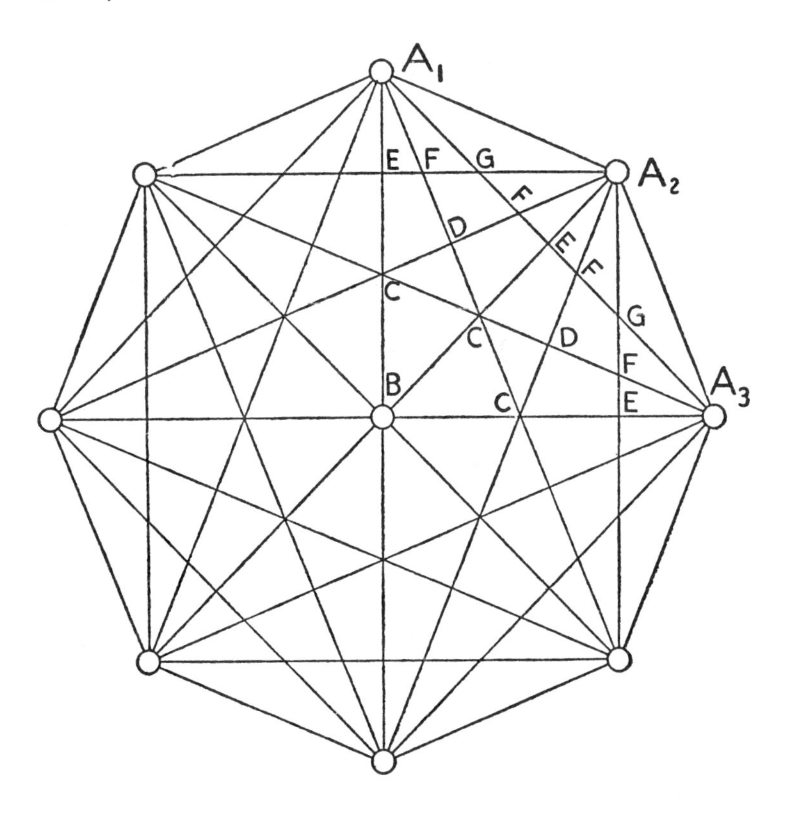

The triangles can be grouped into four classes, according as the number of A points in each is three, two, one, or none.

THREE A POINTS

As every A point is connected with every other, any three A points represent a triangle. The number in this class is thus C^8_3, which is $\frac{8!}{3!\ 5!}=56$.

TWO A POINTS

Here we shall have to take account of four sub-groups, according to the relationship of the two A points. In every case A_1 will be one of the points, so that counting the triangles becomes a matter of counting the available intersections on the available rays through A_1 (some rays and points being excluded by the proviso that one vertex of each triangle must not be of the A class).

(*a*) Leg A_1–A_2. On the other rays through A_1, that is, the rays to A_3, A_4, A_5, A_6, A_7, we find respectively $5+4+3+2+1$ available intersections with rays from A_2, or 15 in all. As there are 8 separate pairs of adjacent A points, the total in this sub-group is $8\times 15=120$.

(*b*) Leg A_1–A_3. On the bundles of rays from each point we find $4+3+2+1$ available intersections, making 10 in all. As there are 8 pairs of alternate A points, the total in this sub-group is $8\times 10=80$.

(*c*) Leg A_1–A_4. On the A_1 rays to A_3, A_5, etc., the available points total $1+3+2+1$, or 7. Total in sub-group, $8\times 7=56$.

(*d*) Leg A_1–A_5. The count of points is $1+2+2+1$, or 6. As there are only 4 pairs of opposite A points, the total in the sub-group is $4\times 6=24$.

For the whole class the total is $120+80+56+24=280$.

ONE A POINT

In the case of A_1, each triangle of this class must have two legs lying on rays from A_1 to A_3, A_4 . . . A_7. If we examine each possible pair of rays, and note every secant that cuts both of the pair within the octagon, we shall take account of all triangles in this class.

(*a*) Rays from A_1 to A_3 and A_4. Here we find 4 triangles. Add 4 more on the rays to A_6, A_7. Sub-group total, $8 \times 8 = 64$.

(*b*) Rays to A_3 and A_5. There are 3 triangles; add 3 for rays A_7, A_5. Sub-group total, $8 \times 6 = 48$.

(*c*) Rays to A_3 and A_6. Two triangles; add the two for A_7, A_4. Total, $8 \times 4 = 32$.

(*d*) Rays to A_3 and A_7. Only one. Total, $8 \times 1 = 8$.

(*e*) Rays to A_4 and A_5. Six triangles, plus 6 for rays A_6, A_5. Total, $8 \times 12 = 96$.

(*f*) Rays to A_4 and A_6. Four triangles. Total, $8 \times 4 = 32$.

We have duly taken account of each of the 10 pairs possible out of the 5 available rays in each bundle. The total for this whole class is 280.

NO A POINT

In the interior octagon formed by the C points, there are 8 triangles having a common vertex at B. Each of the 8 D points is the vertex of a triangle with two C points. These 16 triangles account for all which are wholly interior. None of the E, F, or G points generate interior triangles.

GRAND TOTAL

The whole number of triangles is thus $56+280+280+16 = 632$.

163. THE COIN DROPPER. Not all 14! permutations of the 14 coins are possible. The nature of the dropper prevents the operator from choosing which dime, for instance, he is to take first. He is bound to take the dime at the bottom of the cylinder. How to reckon with this limitation is the real problem.

The solution is easy if we construe the problem as how to fill 14 numbered positions (order of removal) with a batch of 4 nickels, another batch of 3 nickels, a batch of 5 dimes, and 2 quarters. Within each batch we cannot vary the order of the coins, which has been fixed by the order in which they chance to be put into the dropper. Therefore we are concerned only with the combination of position-numbers to be assigned to each batch.

For the batch of 4 nickels, we have a choice of 4 position-numbers out of 14, or $\frac{14!}{4!\,10!}$. For the next batch, say the 3 nickels, 10 remaining positions are open, so that the number of choices is $\frac{10!}{3!\,7!}$. The number of places left for the batch of 5 dimes is $\frac{7!}{5!\,2!}$. There is no choice in the placement of the last batch, 2 quarters. The product of the foregoing fractions is

$$\frac{14!}{5!\,4!\,3!\,2!}=2{,}522{,}520$$

It makes no difference in what order we select the batches to be placed. By cancellation of like terms we always arrive at the above result.

164. ROTATION POOL. There are 16,384 possible orders. The number of ways of filling the rack under the conditions must be the same as the number of ways of removing the 15 balls from the full rack without at any time leaving a gap. In the removal process, there is always choice of the two end balls, neither more nor less, until only one ball remains. The general formula for n balls is therefore 2^{n-1}. For 15 balls, the answer is 2^{14}.

165. THE NECKLACE. No general formula exists which can avoid the necessity for writing out the possible patterns and calculating separately how many different designs each represents. Here is the complete tabulation.

	PATTERN	CHOICE OF COLORS	PERMUTATION OF COLORS	PRODUCT
1 color	AAAAA	4	1	4
2 colors	AAAAB	6	2	12
	AAABB	6	2	12
	ABABB	6	2	12

3 colors	AAABC	4	3	12
	AABAC	4	3	12
	AABBC	4	3	12
	AABCB	4	6	24
	ABCAB	4	3	12
4 colors	AABCD	1	12	12
	ABACD	1	12	12
				136

Each letter of the pattern represents a block of four beads of like color. The patterns must be understood to be circular: the last letter is adjacent to the first. Care must be exercised not to duplicate any patterns, e.g., a form ABBAC would be identical with AABCB, which is listed. Remember that there is no "right and left" in the pattern; you have to reckon both ways in comparing two patterns to see if they are different.

The first column of numbers gives the ways in which 1, 2, 3, or 4 colors can be selected out of 4. The second column gives the ways in which the selected color-names can be permuted against the particular pattern. For example, with the unique pattern AABCB there is choice of any one of 3 colors from which to take the two consecutive blocks (A), then choice of two remaining colors for the single block (C).

166. TOURNAMENT SCHEDULES. The following schedule is made on the same principle as the Howell pair schedules for bridge.

This schedule is "cyclical, central." It is cyclical because the pairings for the first round serve to determine those for the next seven, by a one-step cyclic progression of the players. It is central because one player (9) is placed outside the cyclic chain —or rather, within. A conventional representation is to place this excepted player in the center of a circle, the others being points on the circle. The task of constructing the schedule is essentially to link the circumference points in pairs by chords, no two of which are of the same length. For low numbers of players, this geometrical solution is easy and is in fact the way the schedules are usually constructed.

SCHEDULE FOR NINE PLAYERS

Round	W		B	W		B	W		B	W		B	Bye
1	9	vs.	1	2	vs.	4	3	vs.	6	7	vs.	8	5
2	9		2	3		5	4		7	8		1	6
3	9		3	4		6	5		8	1		2	7
4	9		4	5		7	6		1	2		3	8
5	5		9	6		8	7		2	3		4	1
6	6		9	7		1	8		3	4		5	2
7	7		9	8		2	1		4	5		6	3
8	8		9	1		3	2		5	6		7	4
9	1		5	2		6	3		7	4		8	9

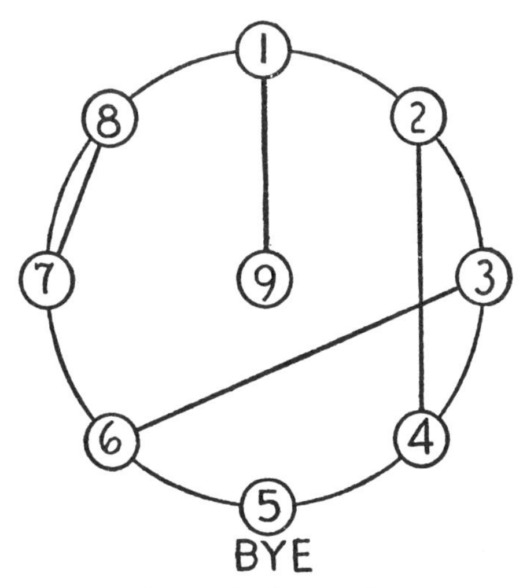

Geometric Construction

Because of the byes, the 9th round is not cyclic with the first eight. Here, all the diametrically opposite points are linked. The matter of giving each player an equal share of the White (W) and the Black (B) sides of the board is easily adjusted. If the columns of the schedule are assigned to positions at four tables, the play at Tables 2, 3, 4 is self-balancing if the side of the White pieces is fixed at each. It remains only to give the central player White four times and Black four times, balancing the distribution against the assignment of colors in the 9th round.

167. PHALANXES. What we have to do is to find the total of all terms of form C^n_r where r takes all values from 1 up to n. In other words, we have to find the total of combinations of n objects taken 1 at a time, 2 at a time, etc., up to $(n-1)$ and then n at a time.

If we write out the terms, they will be of this form:

$$\frac{n!}{n!}+\frac{n!}{(n-1)!}+\frac{n!}{2!(n-2)!}+\frac{n!}{3!(n-3)!}\cdots+\frac{n!}{(n-2)!\,2!}+\frac{n!}{(n-1)!}+\frac{n!}{n!}$$

I do not expect the reader to evaluate this expression by analytical methods. But he can easily determine the sum from data given in this book. If the reader does not at once recognize the above as the formula for the coefficients in the binomial expansion, he will surely recognize it from the calculation for a few low values of n:

for $n=2$, the sum is $1+2+1$
for $n=3$, the sum is $1+3+3+1$
for $n=4$, the sum is $1+4+6+4+1$

Each series appears along a base line of Pascal's triangle (see No. 136—*Figurate Numbers*), and it is stated that these base lines are the coefficients of the binomial expansion.

What is the sum of the series? For $n=2$, the sum is 4; for $n=3$, the sum is 8; for $n=4$, the sum is 16. Evidently the sum is always 2^n. And this in fact is the correct answer to the puzzle: the number of phalanxes under the conditions described in 2^n.

The reader may wonder why, then, the number of phalanxes that can be made from 12 soldiers is 6, which is not a power of 2. The reason is that 12 does not answer to the description of s in the problem: its prime factors are not *all different*. Its factors are $2 \times 2 \times 3$, or $2^2 \times 3$. The general rule, whose derivation is beyond the scope of this book, is that if $a, b, c \ldots$ are different primes, and if $s = (a^p)(b^q)(c^r) \ldots$ then the number of different phalanxes that can be made out of s is

$$(p+1)(q+1)(r+1) \ldots$$

For 12, which equals $(2^2)(3^1)$, the formula gives

$$(2+1)(1+1) = 6$$

169. THAT KING OF CLUBS! Jones is correct in asserting that the odds favor the second finesse. Smith makes a common—and excusable—oversight in the conditions of the problem.

The fundamental assumption from which all other bridge probabilities are derived is that any given hand of 13 cards has the same chance as any other to be dealt. But it does not follow that any hand of 12 or less cards, held after one or more tricks have been played, has the same chance to occur as any other. For the hands are not depleted *at random,* but by willful selection which interferes with the operation of blind chance.

Card probabilities are tabulated as of the original deal. To deduce therefrom the probabilities of distribution in a depleted deal, it is necessary to introduce additional assumptions, based upon the conditions of the game and the habits of the players. Or, to use the term of Pierre Boulanger in this connection, we must reckon with *the probability of causes.*

When South played the Queen on the first Club round, he did indeed exclude Case 1 as a possibility. If Case 2 obtains, it is 100% certain that South played the Queen because he had to. In Case 3, South has a choice of playing Queen or King on the first round. If he is a good player he will vary his procedure, so as to keep his opponents guessing. Let us say that a particular South plays the Queen half the time and the King half the time. Then his play of the Queen excludes the 34 cases where he would have played the King. In other words, the odds are 62 to 34 that he played the Queen because he had to, rather than because he chose to. The conclusion is that the odds for declarer are 62:34 in favor of the second finesse as against swinging the Ace.

The fact that any given South player may not vary his practice just 50-50 does not invalidate the form of this argument. Whatever the figures for South's practice on the play of King or Queen from King-Queen blank, they must be applied to the chances of the original deal to determine *the probability of cause*—the odds that his selection was from choice rather than necessity.

Smith would be right in going up with the Ace if this particular South player were known to play almost invariably the Queen rather than the King. Such practice would of course be very bad, as it would be a dead giveaway in all cases where South held the King without the Queen.

170. ODDS. The correct odds on Calypso are 11 to 4. The odds of 2 to 1 against Agamemnon mean that the probability of his winning is taken to be $\frac{1}{3}$. The probability that Behemoth will win is rated at $\frac{2}{5}$. Calypso's chance is the difference between 1 and the sum $\frac{1}{3}+\frac{2}{5}$, or $\frac{4}{15}$.

Of course in practice the bookmaker must "shorten the odds," to give himself a margin of profit.

171. PARLIAMENT SOLITAIRE. The number of different combinations of 8 cards out of 104 is $\frac{104!}{8!\,96!}$. If we were to count the number of hands holding at least one ace or king, we would have to make a separate computation for one such card, two such cards, etc., then add the results. But if we reckon the combinations of 8 cards that *do not* contain any ace or king, we need deal only with the remaining 88 cards of the deck. The number of such hands is: $\frac{88!}{8!\,80!}$. The probability that the first eight cards dealt *will not* contain ace or king is given by:

$$\frac{88!}{8!\ 80!} \times \frac{8!\ 96!}{104!} = \frac{88!\ 96!}{80!\ 104!}$$

The probability that at least one ace or king *will* turn up is the difference between the above fraction and unity. The actual value of the negative probability above is:

$$\frac{114,575,607}{459,136,405}$$

Subtracting this fraction from 1 gives the positive probability:

$$\frac{344,560,798}{459,136,405}$$

Comparing numerators shows that the chance is about 3 to 1 in favor of success on the first try.

172. EVERY THROW A STRAIGHT. The chance of casting at least one 4 when you throw two dice is not $1/3$ but $11/36$. If you were to write out all 6×6 combinations of numbers that can show on two dice, you would have to write 4 (and each other number) just 12 times, but you could count only 11 different combinations containing a 4, because one of them is 4—4.

As we have seen in No. 171—*Parliament Solitaire,* the probability that an event will occur at least once, where it may occur more than once, is best computed from the *negative probability* that it will not occur at all. What is the chance that you *will not* cast a single 4 with six dice rolled simultaneously? On each die the chance of not casting 4 is $5/6$, and the compound probability on all six together is $5/6 \times 5/6$ . . . or $(5/6)^6$. This fraction of course decreases as the exponent increases. The positive probability of throwing at least one 4 therefore increases with every additional die used. But no matter how many are added, no matter how large the exponent of the negative probability is made, the negative fraction never reaches zero, and the positive probability of casting a 4 never reaches 1 (certainty).

173. TREIZE. I will use the symbol H_n to mean "the number of permutations which give no hit, with n cards." In every case, the total of all permutations is $n!$

For $n=2$, $n!=2$. One permutation (1, 2) hits twice and the other (2, 1) hits not at all. Then $H_2=1$.

For $n=3$, $n!=6$. Suppose that we write out all permutations:

Ordinal number	1 2 3
Permutations	1 2 3
	1 3 2
	*2 3 1
	2 1 3
	*3 1 2
	3 2 1

The permutations starred (*) are the only ones with no hit. Then $H_3=2$.

For $n=4$, $n!=24$. Now instead of writing out all the permutations, let us consider how we *could* write out only the H permutations.

To begin with, we must exclude card No. 1 from the 1st position. That leaves 3, or $n-1$ cards which we can put there:

Ordinal number	1 2 3 4
H *Permutations*	2 . . . . . .
	3 . . . . . .
	4 . . . . . .

Each row must represent several permutations, since there remain three cards to be permuted among three remaining positions. Let us consider each row as representing a sub-group; the number of permutations in all sub-groups must be the same; after we have determined the number per sub-group we can multiply this number by 3 (which is $n-1$) to get the total of H permutations.

How shall we permute the last three numbers in a sub-group? Well, two of the numbers are the same as ordinal positions concerned, while one is different, e.g., on the first row we have to permute 1, 3, 4 among the positions 2, 3, 4. If we treat 1 and 2 for the moment as identical, then the H permutations are H_3, which we have already determined to be 2. But 1 and 2 actually being different, we can place card No. 1 in the second position and get as many additional permutations as we can out of 3, 4 in positions 3, 4. This number is simply H_2, which we have already determined to be 1.

In each sub-group the number of permutations is therefore:

$$H_{n-1}+H_{n-2}$$

As there are $n-1$ sub-groups, we have finally:

$$H_n=(n-1)(H_{n-1}+H_{n-2})$$

This is the formula whereby we can go on compounding the table of H permutations as n increases. To solve the given problem we will build the table up to $n=6$:

n	H_n	$n!$	$H_n/n!$
2	1	2	·5
3	2	6	·33··
4	9	24	·375
5	44	120	.366··
6	265	720	.368

The chances of winning the Treize patience with six cards are thus about 10 in 27. If the reader cares to carry the table a little further, he will discover that as n increases the ratio in the rightmost column remains stable at .367+.

176. TO LEAVE THE LAST. A player can be forced to take the last counter only when there is but one left. Consequently the formula is

$$w=1+(a+m)n$$

177. TO WIN THE ODD. The formula can be found by mathematical induction. Fix limits for the draw, as 1 to 3, then plumb every possibility when the common pile is reduced to 4, 5, etc. It will be found that the outcome depends in every case on whether the player has at that juncture acquired an odd or an even number of stones.

Against a player who has acquired an *odd* number:

$$w=2n(a+m)+a[+m]$$

For each set of values given to a, m, n, the formula gives two values for w, the square brackets indicating that in one case the term is to be added and in the other it is to be omitted.

Against a player who owns an *even* number:

$$w=2n(a+m)[+2a+m]$$

In the case of $a=1$ and $m=3$, these formulas reduce to

Against *odd,* the series $1+8n$ (1,9,17. .) and $4+8n$ (4,12,20. . .) Against *even,* the series $8n$ (8,16,24. . .) and $5+8n$ (5,13,21. . .)

178. THREE-FIVE-SEVEN. The powers of 2 (2^0, 2^1, 2^2, etc.) are the numbers 1, 2, 4, etc. Every integer can be expressed as the sum of a group of these numbers *in a unique way.* In other words, if each number of the series is used once only, there is only one way of choosing the right numbers to sum to a given integer.

For example, the integer 59 can be broken down into 32+16+8+2+1. The way to discover the unique series for each integer is to subtract from it and from each remainder thereafter the largest possible power of 2.

In the game of three piles, express the number of counters in each pile in powers of 2. The initial array, for example, is

1	1	1
2	4	2
		4
3	5	7

The law involved is that an array is a w (a winner for the player who presents it to his opponent) if each power of 2 therein represented at all appears *just twice.*

In the initial array, there are two 4's and two 2's, but three 1's. The first player wins by taking one counter from any pile.

It is evident that a player who has to draw from two equal piles is lost. In drawing, therefore, each player must be careful not to exhaust any pile unless the other two are equal.

This solution is quite general; it can be applied to any number of piles and any number of counters. The formula is then to be read that an array is w when each power of 2 represented at all appears *an even number of times.*

Notice that it is not always possible to reduce an array to w merely by subtracting some of its component numbers. Occasion will arise when it is necessary to subtract a number not visible in the array, so as to change one series of powers to another. For example, with three piles of a larger number of counters, suppose this array is reached:

2	1	2
4	2	4
	4	8
6	7	14

To take 6 counters from either of the larger piles will equalize the 2's and 4's, but will leave the 1 and 8 unmatched. The solution is to draw 13 from the largest pile, leaving the array

2	1	1
4	2	
	4	
6	7	1

If the separate totals of the piles are expressed in the binary system (where there are only two digits instead of ten), the determination of w is a simple matter of subtraction among these numbers. The solution as given actually depends on this fact, since it expresses the integers of the binary scale in the decimal system to which we are accustomed.

179. THE THIRTY-ONE GAME. As pointed out by Dudeney, the simplest course for the first player is to turn 5. If the second turns 5 to get into the w series, the first player turns 2. If the second then persists in the series, first continues with 2's and makes 26 with the 5's exhausted. But if the second player at any time goes out of the *w* series, first can seize it and no needed number can be exhausted.

The first player can also win, by more complicated play, after commencing with A or 2.

180. THIRTY-ONE WITH DICE. How to attack this problem is puzzling until one hits on the idea of digital roots.

The digital root of the goal 31 is 4. The *w* numbers (totals that win for the player who reaches them) are primarily those whose digital roots are 4, namely 4, 13, 22, 31.

The members of this primary series differ by 9. If the player confronted by a *w* adds 1 or 2, his opponent at the next turn cannot reach the next *w* number. How is the win to be enforced?

Well, the player who cannot reach a *w* should do the next best thing—he should *prevent his opponent* from getting into the series. To effect this he must reach a root which differs from root 4 by the *number he turns up or its complement.* These two numbers, remember, are not available to the next player.

When 1 or 2 is added to root 4, the root becomes 5 or 6. Then the addition of 4 or 3 respectively to reach 9 holds the win, since in either case 4 is unavailable.

Evidently we must add to the primary *w* series: any total with root 9 reached by turning 3 or 4.

Suppose that to root 9 a player adds 2, making root 2. Then the next player cannot reach root 4, because the 2 is unavailable; nor root 9, because there is no 7 on the die. What is he to do? He must prevent the other from reaching any *w,* 4 or 9. The only play is to turn 3, making root 5, and then the other cannot reach 9 because the complementary 4 is buried.

By similar reasoning, it can be shown that, besides the primary *w* series of root 4, there exist also conditional *w* numbers as follows:

Root 1, 5, or 9 reached by turning 3 or 4;

Root 8 reached by turning 2 or 5.

All the other roots, 2, 3, 6, 7, are sure losers. A number will always be available to reach a root of the *w* series: for root 2, either 2 or 3 will serve; for 3, 1 or 5; for 6, 2 or 3; for 7, 3 or 6.

The only roll that assures the first player a win is of course 4.

182. SAM LOYD'S DAISY PUZZLE. The second player wins. The method is to break the circle of petals into two symmetrical semicircles. Whether the first player takes 1 petal or 2, the second player must answer by taking 2 or 1 respectively from the opposite end of the diameter, breaking the array into two blocks of 5 petals each. The second player then continues to restore the symmetry each time it is broken by the first, and consequently must draw the last petal.

183. DUDENEY'S CIGAR PUZZLE. After all we have said of the principle of symmetry, the reader will no doubt perceive the idea of this puzzle. Whoever has to play from a perfectly symmetrical arrangement of cigars on the table must lose. If he can find room to add a cigar, then so can his opponent. The latter need only add his cigar on that point (and in same relative orientation) which is symmetrical to the position of the first cigar, with respect to the center point of the table. (Two points A and B are symmetrical with respect to a point C, if C is the midpoint of the straight line joining A and B.)

It would seem then that the second player must win, since he has only to keep pairing his plays with those of the first player. But there is a circumstance easily overlooked. The cigar as described can be made to stand upright on its flat end. The first player wins by placing his first cigar upright on the only unique point of the table—the center of symmetry itself!

184. THE CARPATHIAN SPIDER. At the outset the fly stands on the same line as the spider, 8 points away (an even number). That means that whenever the two insects stand on diagonally-opposite points of the same quadrilateral, it will be the spider's turn to move. If the whole web were composed of quadrilaterals, the fly could never be caught—he could merely circulate around the four sides of a quadrilateral.

But the web includes one triangle. The spider need only go at once to this triangle and circumnavigate it to change "the move." Thereafter it will be the fly's turn to move when the number of points intervening between the two insects is even. A little experiment will show that the fly is then easily run down.

The puzzle is seen to be another illustration of the principle of symmetry. The opposition play has its analogue in chess, where the maneuver of "triangulation" is well-known as a necessity to change the move in some end-situations.

The "spider and fly" puzzle is here presented in its simplest form. The principle is capable of great elaboration through the addition of other ideas: pentagonal and heptagonal figures as well as triangles; more than one circuit of an odd number of points in the web; opportunities for both players to change the move but a limitation on the opportunity of one to reach an odd circuit without being cornered there; arbitrary rules such as limitation of the number of times some specified intersections may be visited. The whole subject presents a wealth of opportunities to those who like to make puzzles—and this is the type of puzzle that is as much fun to make as to solve!

189. SALVO. As with any question of probability, the answer can be given only in terms of some basic assumption. Here we assume that the player, having resolved to fire his first salvo in the given formation, is just as likely to place it in one orientation as in any other. We can sum up the total number of times each square is hit if every possible orientation is used once, and this sum is the measure of the probability that the particular square will be hit.

An easy method of operation is as follows:

From a piece of paper cut out 7 squares in the chosen formation. Using this as a mask or "grille," lay it on a 10×10 square, move it vertically and horizontally (without rotation) to every possible position within the square, and in each position mark a tick through each of the 7 holes onto the paper below. The result of this step is shown in Fig. 1.

Fig. 1

2	3	4	5	4	4	5	4	3	2
3	4	7	11	11	11	11	7	4	3
4	7	12	17	18	18	17	12	7	4
5	11	17	22	25	25	22	17	11	5
4	11	18	25	30	30	25	18	11	4
4	11	18	25	30	30	25	18	11	4
5	11	17	22	25	25	22	17	11	5
4	7	12	17	18	18	17	12	7	4
3	4	7	11	11	11	11	7	4	3
2	3	4	5	4	4	5	4	3	2

Fig. 2

The formation is not symmetrical. We have to take account of three other possible orientations corresponding to rotations of the grille through 90, 180, and 270 degrees. To do so, merely make three copies of the diagram, then align all four side by side in the four relative rotations. In a new square, sum the ticks in corresponding cells of all four squares.

Finally, we have to take account of four more orientations which are equivalent to turning the grille over. Make a copy of the summation square in mirror reversal, then sum it with the copy in yet another square. The final result, Fig. 2, shows how many times each cell is hit if every possible orientation is used once.

From Fig. 2 we see that the safest position for the battleship is in a corner, extending vertically or horizontally. The worst position is vertically or horizontally across the middle of the arena. The relative chances of being hit sum to 18 and 128 respectively, or about 7 to 1 in favor of the corner.

Appendix

HOW TO EXTRACT SQUARE ROOT. For illustration let us extract the square root of 4,375,690,201. The working sheet is shown below, and the operations are described step by step.

```
    6  6  1  4  9
  ________________
 √43 75 69 02 01
  36
  --
   7 75                        120
   7 56                        126
   ----
     19 69                     1320
     13 21                     1321
     -----
      6 48 02                  13220
      5 28 96                  13224
      --------
      1 19 06 01               132280
      1 19 06 01               132289
      ----------
```

1. Divide the number into periods of two digits each, from right to left.

2. Below the first period (leftmost) write the largest square that is equal to it or less (36), and write its square root (6) as the first digit of the quotient.

3. Subtract the square from the first period, then bring down the next period, to make the first remainder (775).

4. In a separate memorandum column to the right, double the quotient (so far as it goes) and add one zero (120).

5. Now estimate the digit that must be added to this memo number so that, when the total is multiplied by the same digit, the product will be the largest possible equal to or less than the first remainder. (120 goes into 775, at most, 6 times. When 6 is added to 120, the sum 126 multiplied by 6 does not exceed 775, so that 6 is the correct second digit.) Write this digit, when correctly estimated, as the second digit of the quotient.

6. Write the new product just found (756) under the first remainder, subtract, and bring down the next period to make the second remainder.

7. Repeat steps 4, 5, 6 until all the periods are exhausted.

Take note that the memo number written at the right must always be double the *entire* quotient, so far as it is determined, plus *one zero for each period brought down* into the remainder. If a remainder should be smaller than the memo number, then an additional period must be brought down and another zero affixed to the memo number.

It is worthwhile to understand the why and wherefore of these operations.

The square root sought is construed to be of binomial form, $a+b$. Its square is consequently of form $a^2+2ab+b^2$. The quantity a is the integral part of the root, while b is any remainder. The part a is found by successive approximations. Thus, when we wrote 6 as the first digit of the root in the above example we were really estimating that the root lies between 60,000 and 70,000. Our first approximation 60,000 being a, we subtracted a^2 from the original number, leaving 775,690,201. This difference is in form $2ab+b^2$, which we treat as $b(2a+b)$. We wrote $2a$ in the memo column: this number is really 120,000, but we leave off the last three zeros because we have left the last six figures off the first remainder. We next have to estimate an integer b which will make $b(2a+b)$ the largest possible partial product. Having done so, we write this digit into the root, which now is approximated more closely to 66,000. This new value becomes a new a, and we proceed to estimate a new b that will give the third digit of the root, and so on.

HOW TO EXTRACT CUBE ROOT. Before studying this operation, read the preceding algebraic explanation of the extraction of square root.

The operations in finding a cube root are based on similar principles of algebra. The root is construed as a polynomial, $a+b$, an integral part plus a remainder. The cube is consequently of form $a^3+3a^2b+3ab^2+b^3$. The part a is found by successive approximations. To illustrate:

```
      4   1   3
  ∛70 444 997
   64
   --
    6 444            4800
    4 921             120
    -----            ----
    1 523 997           1          504300
    1 523 997        4921            3690
    ---------                           9
                                   ------
                                   507999
```

1. Divide the number into periods of three digits each, from right to left.

2. Below the first period (leftmost) write the largest cube that is equal to it or less (64), and write its cube root (4) as the first digit of the quotient.

3. Subtract the cube from the first period, then bring down the next period, to make the first remainder (6444).

4. In a separate memorandum column to the right, enter three times the square of the quotient (so far as it goes) and add two zeros (4800, which is $3 \times 4^2 \times 100$).

The quotient so far as it goes is a; the remainder after a^3 is subtracted is of form $3a^2b+3ab^2+b^3$, which is treated as $b(3a^2+3ab+b^2)$. We have written $3a^2$ (4800); now we must find a digit b so that to 4800 we can add $120b$ and b^2, then multiply the sum by b to produce the largest possible partial product.

Since 4800 will go into 6444 only once, we see that the b is here 1, so we write 1 as the second digit of the root, add 120 and 1 to 4800, and subtract the sum from 6444.

5. Repeat steps 4 and 5 until all the periods are exhausted.

Take note that in writing $3a^2$ in the memo column it is necessary to *add two zeros for each period* brought down into the remainder, and in writing $3a$ (to be made into $3ab$ when b is estimated) it is necessary to *add one zero for each period* brought down.

If you become confused about the zeros, just picture each line of figures in the computation extended by zeros to the right as far as the rightmost column. The first remainder 6444 is thus seen to be really 6,444,000, and the first digit of the root 4 really stands for 400 (one digit for each period). Now if a is 400, then $3a^2$ is 480,000; this is going to be multiplied by a digit b which really is $10b$, since it will stand in the tens place of the root. The product will then be of the order 4,800,000. Since for convenience we omit the last three zeros from 6,444,000, we must likewise omit just three terminal zeros from 4,800,000 if we are to align this number at the right below the first remainder.

Similarly, $3a$ is actually 1200, and since this is to be multiplied by b^2 (which is actually $100b^2$) the product will be of order 120,000. Finally the digit b is going to be cubed, so it will be of order $1{,}000b$.

TABLE OF SQUARE ROOTS

2—1.41421
3—1.73205
5—2.2360
6—2.4494
7—2.6457
8—2.8284
10—3.1622
11—3.3166

TABLE OF POWERS OF 2

1st —2
2nd—4
3rd—8
4th—16
5th—32
6th—64
7th—128
8th—256
9th—512
10th—1024
11th—2048
12th—4096
13th—8192
14th—16,384
15th—32,768
16th—65,536
17th—131,072
18th—262,144
19th—524,288
20th—1,048,576

TABLE OF SQUARE NUMBERS

1—1
2—4
3—9
4—16
5—25

6—36
7—49
8—64
9—81
10—100

11—121
12—144
13—169
14—196
15—225

16—256
17—289
18—324
19—361
20—400

21—441
22—484
23—529
24—576
25—625

26—676
27—729
28—784
29—841
30—900

31—961
32—1024
33—1089
34—1156
35—1225

36—1296
37—1369
38—1444
39—1521
40—1600

41—1681
42—1764
43—1849
44—1936
45—2025

46—2116
47—2209
48—2304
49—2401
50—2500

51—2601
52—2704
53—2809
54—2916
55—3025

56—3136
57—3249
58—3364
59—3481
60—3600

61—3721
62—3844
63—3969
64—4096
65—4225

66—4356
67—4489
68—4624
69—4761
70—4900

71—5041
72—5184
73—5329
74—5476
75—5625

76—5776
77—5929
78—6084
79—6241
80—6400

81—6561
82—6724
83—6889
84—7056
85—7225

86—7396
87—7569
88—7744
89—7921
90—8100

91—8281
92—8464
93—8649
94—8836
95—9025

96—9216
97—9409
98—9604
99—9801
100—10,000

101—10,201
102—10,404
103—10,609
104—10,816
105—11,025

106—11,236
107—11,449
108—11,664
109—11,881
110—12,100

111—12,321
112—12,544
113—12,769
114—12,996
115—13,225

116—13,456
117—13,689
118—13,924
119—14,161
120—14,400

TABLE OF PRIME NUMBERS

(between 1 and 1,000)

2
3
5
7

11
13
17
19
23
29
31
37
41
43
47
53
59
61
67
71
73
79
83
89
97

101
103
107
109
113
127
131
137
139
149
151
157
163
167
173
179
181

191
193
197
199

211
223
227
229
233
239
241
251
257
263
269
271
277
281
283
293

307
311
313
317
331
337
347
349
353
359
367
373
379
383
389
397

401
409
419
421
431

433
439
443
449
457
461
463
467
479
487
491
499

503
509
521
523
541
547
557
563
569
571
577
587
593
599

601
607
613
617
619
631
641
643
647
653
659
661
673
677
683
691

701
709
719
727
733
739
743
751
757
761
769
773
787
797

809
811
821
823
827
829
839
853
857
859
863
877
881
883
887

907
911
919
929
937
941
947
953
967
971
977
983
991
997

TABLE OF TRIANGULAR NUMBERS

1—1
2—3
3—6
4—10
5—15

6—21
7—28
8—36
9—45
10—55

11—66
12—78
13—91
14—105
15—120

16—136
17—153
18—171
19—190
20—210

21—231
22—253
23—276
24—300
25—325

26—351
27—378
28—406
29—435
30—465

31—496
32—528
33—561
34—595
35—630

36—666
37—703
38—741
39—780
40—820

41—861
42—903
43—946
44—990
45—1035

46—1081
47—1128
48—1176
49—1225
50—1275

51—1326
52—1378
53—1431
54—1485
55—1540

56—1596
57—1653
58—1711
59—1770
60—1830

Glossary

additive—Any one of several numbers that are added together.

azimuth—The angle between a line of direction and north, measured clockwise from north.

binomial theorem—The formula for determining the coefficients of the expansion of $(a+b)^n$ for any value of n.

Caliban puzzle—One in which the solver is asked to infer a fact from a set of given facts, so-called from the pseudonym of an English inventor of such puzzles.

cardinal number—One expressing magnitude or quantity, as: 2, 561, a million.

coefficient—A number (or symbol) prefixed to another number and by which the latter is multiplied, as: 2 in $2x=y$; A, B, C in $Ax^2+Bx+C=O$.

combination—Any particular sub-group selected from a larger group, as: the combination BD out of ABCDE.

combinatorial analysis—The study of combinations, especially as to kinds and classes as well as to number.

complex number—Imaginary number; an expression involving both real and imaginary numbers.

composite number—One that has some factors besides itself and unity, as: 6 $(=3\times2)$, 385 $(=5\times7\times11)$. *Antonym,* prime number.

constant—A number of fixed value, as: 3, $\sqrt{5}$, $\sqrt[3]{8}$, $13!$, π. *Antonym,* variable.

cube root—Any one of the three equal factors whose product is the given number as: 2 is the cube root of 8.

cubic equation—One in which some variable has the exponent 3, but no higher exponent appears, as $4x^3=7y-15$.

digit—Any of the nine symbols 1 2 3 4 5 6 7 8 9, possibly also including the symbol 0, zero.

digital root—The sum of the digits of an integer, continued to a single digit, as: the digital root of 786 is 3 ($7+8+6=21$ and $2+1=3$).

Diophantine equation—One of indeterminate form but with limited solutions by reason of the fact that the values of the variables must be integral.

discrete—Discontinuous.

empiric—Based upon experience (as trial and error) rather than on theoretic formula.

expansion—The carrying out of an indicated operation, as multiplication or the raising to a power.

explicit equation—One that states the value of a variable in terms of another or others, as $x=y$, $3y=5$. *Antonym,* implicit equation.

exponent—An operator written as a superscript, indicating the number of times the number to which it is attached is to be used as a factor, as: $3^5=243$ ($3\times3\times3\times3\times3$).

factor—A number by which another can be evenly divided; any one of the terms in the expression of a product.

factorial (of an integer)—The product of all integers from $1\times2\times3$ . . . up to and including the given integer, as: factorial $4=24$ (the product $1\times2\times3\times4$); the term "factorial" is expressed by the symbol ! as in $n!$ or by $\underline{\ \ }\!\rfloor$ as in $\underline{n}\!\rfloor$.

factorization—The determination of the factors of a number.

G.C.D.—Means "greatest common divisor"; the largest number that is a factor of each of several given numbers.

higher arithmetic—The study of certain properties of integers, now usually called theory of numbers.

hypotenuse—The longest side of a right triangle.

imaginary number—The square root of —1, or any multiple thereof; the symbol i is used to denote $\sqrt{-1}$. *Antonym,* real number.

implicit equation—One in which the value of the variables is given indirectly by a relation among them, as: $x+xy+5y=11$. *Antonym,* explicit equation.

integer—A whole number as: 4, 15, 9062.

irrational number—One that cannot be expressed as the ratio between two integers, as the value of π.

L.C.M.—Means "least common multiple"; the smallest number that contains as a factor each of several given numbers.

leg—Either side of a right triangle other than the hypotenuse.

linear equation—One in which no variable has an exponent greater than unity.

multiple—Converse of factor; A is a multiple of B if B is a factor of A.

negative number—One to which the minus sign is prefixed. *Antonym,* positive number.

number—The general term for the symbols of mathematics, as: the digits, the integers, rational and real numbers, imaginary numbers, symbols denoting constants or variables.

numbers, theory of—The study of classes and properties of integers.

odds—A way of stating probability, the ratio of the probability that an event will occur to the probability that it will not.

operator—A number indicating an operation to be performed on other numbers to which it is attached, as: an exponent, the symbol $\surd$.

ordinal number—One that expresses position within a series, as: second, 561st, millionth. *Antonym,* cardinal number.

partial product—In long multiplication, any one of the results of the multiplicand times a digit of the multiplier.

pencil—An aggregate of lines all intersecting at one point.

perfect number—One that is the sum of all its divisors, as 6 $(=1+2+3)$.

permutation—Any one of the arrangements or order in which a group of objects may be placed, as: the permutations ABCDE, BDCAE, ECDAB.

polynomial—An algebraic expression of the sums and/or differences among two or more terms, as $a+b$.

positive number—One to which the plus sign is prefixed; when no sign is attached, a number is understood to be positive. *Antonym,* negative number.

power—Converse of root; A is a power of B if B is a root of A.

prime number—One that has no divisors other than itself and unity, as: 3, 7, 19, 31. *Antonym,* composite number.

probability—The likelihood of occurrence of an event in a certain way, especially when capable of expression in mathematical terms.

proportion—An equality between two ratios, as: $6:4=3:2$; $a:b=c:d$.

Pythagorean theorem—The proposition that $a^2+b^2=c^2$ where a and b are the legs of a right triangle and c is the hypotenuse.

quadratic equation—One in which some variable has the exponent 2, but no higher exponent appears, as: $3x^2+4=2y-5$.

quartic equation—One in which some variable has the exponent 4, but no higher exponent appears, as: $y=x^4+2x^3+5x^2+9x+13$.

radical sign—The symbol $\sqrt[3]{\ }$, expressing the extraction of a root; the degree of the root is written as a superscript in the $\sqrt{\ }$, except that 2 is omitted in expressing square root.

radix—The base of a system of enumeration, as 10, the base of modern numerals.

ratio—The quotient of two numbers, as: 5:4, read "five to four" and denoting $\frac{5}{4}$.

rational number—One that can be expressed as the ratio of two integers.

real number—The class of all rational and irrational numbers together. *Antonym,* imaginary number.

root—Any one of the factors, all equal, whose product is the given number; the *degree* of the root is the number of such factors, as: 3 is the 4th root of 81.

square number—One that is the product of two equal factors, as: 25 (5×5); 381 (19×19).

square root—One of the two equal factors of which the given number is the product, as: 3 is the square root of 9.

symbols—

$+$	plus sign, read "plus"
$-$	minus sign, read "minus"
$\times$ or $\cdot$	multiplication sign, read "times"
$\div$	division sign, read "divided by"
$<$	read "is less than," as $3<4$, 3 is less than 4
$>$	read "is greater than," as $p>q$, p is greater than q
$:$	read "to," indicates ratio, as 5:4, denoting $\frac{5}{4}$
$=$	equality sign, read "equals"
$\neq$	read "does not equal"
$\pm$	read "plus or minus"

7^2 exponent, written as superscript, 2 is read "squared," 3 is read "cubed," 4 and higher is read "to the 4th power," "to the nth power"; indicates the number of times the quantity to which it is affixed is to be used as a factor, as $7^2=7\times7=49$

$\sqrt[3]{\ }$ radical sign, indicates extraction of a root, whose degree is indicated by the number written as a superscript in the $\sqrt{\ }$; without superscript the radical sign is understood to mean square root; 3rd degree is read "cube root," 4th and higher degrees are read "4th root," "nth root"

p_3 subscript, read "sub 3," is used to distinguish separate members of a group when the same symbol (as p) is used to denote all members

! or $\underline{\ \ }|$ factorial sign, read "factorial n" or "n factorial," denotes the product $1\times2\times3 \ldots n$, as $5!$ or $\underline{5}|$ $=1\times2\times3\times4\times5=120$

π pi, read "pi," the constant 3.14159 . . . , the ratio of the circumference of a circle to the diameter

P read "the number of permutations of"

C read "the number of combinations of"

triangular number—One that is the sum of all consecutive integers from unity up to a given integer, as: the triangle of $5=1+2+3+4+5=15$.

variable—A symbol used to represent a quantity of unknown or variable magnitude, as x, p_n.

vertex—The point of intersection of two adjacent sides of a plane figure, as a corner of a square.

A CATALOGUE OF SELECTED DOVER BOOKS IN ALL FIELDS OF INTEREST

A CATALOGUE OF SELECTED DOVER BOOKS IN ALL FIELDS OF INTEREST

150 MASTERPIECES OF DRAWING, edited by Anthony Toney. 150 plates, early 15th century to end of 18th century; Rembrandt, Michelangelo, Dürer, Fragonard, Watteau, Wouwerman, many others. 150pp. 8⅜ x 11¼. 21032-4 Pa. **$4.00**

THE GOLDEN AGE OF THE POSTER, Hayward and Blanche Cirker. 70 extraordinary posters in full colors, from Maîtres de l'Affiche, Mucha, Lautrec, Bradley, Cheret, Beardsley, many others. 9⅜ x 12¼. 22753-7 Pa. **$5.95**
21718-3 Clothbd. $7.95

SIMPLICISSIMUS, selection, translations and text by Stanley Appelbaum. 180 satirical drawings, 16 in full color, from the famous German weekly magazine in the years 1896 to 1926. 24 artists included: Grosz, Kley, Pascin, Kubin, Kollwitz, plus Heine, Thöny, Bruno Paul, others. 172pp. 8½ x 12¼. 23098-8 Pa. $5.00
23099-6 Clothbd. $10.00

THE EARLY WORK OF AUBREY BEARDSLEY, Aubrey Beardsley. 157 plates, 2 in color: Manon Lescaut, Madame Bovary, Morte d'Arthur, Salome, other. Introduction by H. Marillier. 175pp. 8½ x 11. 21816-3 Pa. $4.00

THE LATER WORK OF AUBREY BEARDSLEY, Aubrey Beardsley. Exotic masterpieces of full maturity: Venus and Tannhäuser, Lysistrata, Rape of the Lock, Volpone, Savoy material, etc. 174 plates, 2 in color. 176pp. 8½ x 11. 21817-1 Pa. **$4.00**

DRAWINGS OF WILLIAM BLAKE, William Blake. 92 plates from Book of Job, Divine Comedy, Paradise Lost, visionary heads, mythological figures, Laocoön, etc. Selection, introduction, commentary by Sir Geoffrey Keynes. 178pp. 8½ x 11.
22303-5 Pa. **$4.00**

LONDON: A PILGRIMAGE, Gustave Doré, Blanchard Jerrold. Squalor, riches, misery, beauty of mid-Victorian metropolis; 55 wonderful plates, 125 other illustrations, full social, cultural text by Jerrold. 191pp. of text. 8⅛ x 11.
22306-X Pa. $6.00

THE COMPLETE WOODCUTS OF ALBRECHT DÜRER, edited by Dr. W. Kurth. 346 in all: Old Testament, St. Jerome, Passion, Life of Virgin, Apocalypse, many others. Introduction by Campbell Dodgson. 285pp. 8½ x 12¼. 21097-9 Pa. $6.00

THE DISASTERS OF WAR, Francisco Goya. 83 etchings record horrors of Napoleonic wars in Spain and war in general. Reprint of 1st edition, plus 3 additional plates. Introduction by Philip Hofer. 97pp. 9⅜ x 8¼. 21872-4 Pa. $3.50

ENGRAVINGS OF HOGARTH, William Hogarth. 101 of Hogarth's greatest works: Rake's Progress, Harlot's Progress, Illustrations for Hudibras, Midnight Modern Conversation, Before and After, Beer Street and Gin Lane, many more. Full commentary. 256pp. 11 x 14. 22479-1 Pa. **$7.95**

PRIMITIVE ART, Franz Boas. Great anthropologist on ceramics, textiles, wood, stone, metal, etc.; patterns, technology, symbols, styles. All areas, but fullest on Northwest Coast Indians. 350 illustrations. 378pp. 20025-6 Pa. $3.75

MOTHER GOOSE'S MELODIES. Facsimile of fabulously rare Munroe and Francis "copyright 1833" Boston edition. Familiar and unusual rhymes, wonderful old woodcut illustrations. Edited by E.F. Bleiler. 128pp. 4½ x 6⅜. 22577-1 Pa. $1.50

MOTHER GOOSE IN HIEROGLYPHICS. Favorite nursery rhymes presented in rebus form for children. Fascinating 1849 edition reproduced in toto, with key. Introduction by E.F. Bleiler. About 400 woodcuts. 64pp. 6⅞ x 5¼. 20745-5 Pa. $1.50

PETER PIPER'S PRACTICAL PRINCIPLES OF PLAIN & PERFECT PRONUNCIATION. Alliterative jingles and tongue-twisters. Reproduction in full of 1830 first American edition. 25 spirited woodcuts. 32pp. 4½ x 6⅜. 22560-7 Pa. $1.25

THE NIGHT BEFORE CHRISTMAS, Clement Moore. Full text, and woodcuts from original 1848 book. Also critical, historical material. 19 illustrations. 40pp. 4⅝ x 6. 22797-9 Pa. $1.35

THE KING OF THE GOLDEN RIVER, John Ruskin. Victorian children's classic of three brothers, their attempts to reach the Golden River, what becomes of them. Facsimile of original 1889 edition. 22 illustrations. 56pp. 4⅝ x 6⅜. 20066-3 Pa. $1.50

DREAMS OF THE RAREBIT FIEND, Winsor McCay. Pioneer cartoon strip, unexcelled for beauty, imagination, in 60 full sequences. Incredible technical virtuosity, wonderful visual wit. Historical introduction. 62pp. 8⅜ x 11¼. 21347-1 Pa. $2.50

THE KATZENJAMMER KIDS, Rudolf Dirks. In full color, 14 strips from 1906-7; full of imagination, characteristic humor. Classic of great historical importance. Introduction by August Derleth. 32pp. 9¼ x 12¼. 23005-8 Pa. $2.00

LITTLE ORPHAN ANNIE AND LITTLE ORPHAN ANNIE IN COSMIC CITY, Harold Gray. Two great sequences from the early strips: our curly-haired heroine defends the Warbucks' financial empire and, then, takes on meanie Phineas P. Pinchpenny. Leapin' lizards! 178pp. 6⅛ x 8⅜. 23107-0 Pa. $2.00

WHEN A FELLER NEEDS A FRIEND, Clare Briggs. 122 cartoons by one of the greatest newspaper cartoonists of the early 20th century — about growing up, making a living, family life, daily frustrations and occasional triumphs. 121pp. 8½ x 9½. 23148-8 Pa. $2.50

ABSOLUTELY MAD INVENTIONS, A.E. Brown, H.A. Jeffcott. Hilarious, useless, or merely absurd inventions all granted patents by the U.S. Patent Office. Edible tie pin, mechanical hat tipper, etc. 57 illustrations. 125pp. 22596-8 Pa. $1.50

THE DEVIL'S DICTIONARY, Ambrose Bierce. Barbed, bitter, brilliant witticisms in the form of a dictionary. Best, most ferocious satire America has produced. 145pp. 20487-1 Pa. $1.75

THE BEST DR. THORNDYKE DETECTIVE STORIES, R. Austin Freeman. The Case of Oscar Brodski, The Moabite Cipher, and 5 other favorites featuring the great scientific detective, plus his long-believed-lost first adventure — 31 New Inn — reprinted here for the first time. Edited by E.F. Bleiler. USO 20388-3 Pa. $3.00

BEST "THINKING MACHINE" DETECTIVE STORIES, Jacques Futrelle. The Problem of Cell 13 and 11 other stories about Prof. Augustus S.F.X. Van Dusen, including two "lost" stories. First reprinting of several. Edited by E.F. Bleiler. 241pp. 20537-1 Pa. $3.00

UNCLE SILAS, J. Sheridan LeFanu. Victorian Gothic mystery novel, considered by many best of period, even better than Collins or Dickens. Wonderful psychological terror. Introduction by Frederick Shroyer. 436pp. 21715-9 Pa. $4.00

BEST DR. POGGIOLI DETECTIVE STORIES, T.S. Stribling. 15 best stories from EQMM and The Saint offer new adventures in Mexico, Florida, Tennessee hills as Poggioli unravels mysteries and combats Count Jalacki. 217pp. 23227-1 Pa. $3.00

EIGHT DIME NOVELS, selected with an introduction by E.F. Bleiler. Adventures of Old King Brady, Frank James, Nick Carter, Deadwood Dick, Buffalo Bill, The Steam Man, Frank Merriwell, and Horatio Alger — 1877 to 1905. Important, entertaining popular literature in facsimile reprint, with original covers. 190pp. 9 x 12. 22975-0 Pa. $3.50

ALICE'S ADVENTURES UNDER GROUND, Lewis Carroll. Facsimile of ms. Carroll gave Alice Liddell in 1864. Different in many ways from final Alice. Handlettered, illustrated by Carroll. Introduction by Martin Gardner. 128pp. 21482-6 Pa. $2.00

ALICE IN WONDERLAND COLORING BOOK, Lewis Carroll. Pictures by John Tenniel. Large-size versions of the famous illustrations of Alice, Cheshire Cat, Mad Hatter and all the others, waiting for your crayons. Abridged text. 36 illustrations. 64pp. 8¼ x 11. 22853-3 Pa. $1.50

AVENTURES D'ALICE AU PAYS DES MERVEILLES, Lewis Carroll. Bué's translation of "Alice" into French, supervised by Carroll himself. Novel way to learn language. (No English text.) 42 Tenniel illustrations. 196pp. 22836-3 Pa. $3.00

MYTHS AND FOLK TALES OF IRELAND, Jeremiah Curtin. 11 stories that are Irish versions of European fairy tales and 9 stories from the Fenian cycle — 20 tales of legend and magic that comprise an essential work in the history of folklore. 256pp. 22430-9 Pa. $3.00

EAST O' THE SUN AND WEST O' THE MOON, George W. Dasent. Only full edition of favorite, wonderful Norwegian fairytales — Why the Sea is Salt, Boots and the Troll, etc. — with 77 illustrations by Kittelsen & Werenskiöld. 418pp. 22521-6 Pa. $4.50

PERRAULT'S FAIRY TALES, Charles Perrault and Gustave Doré. Original versions of Cinderella, Sleeping Beauty, Little Red Riding Hood, etc. in best translation, with 34 wonderful illustrations by Gustave Doré. 117pp. 8⅛ x 11. 22311-6 Pa. $2.50

EARLY NEW ENGLAND GRAVESTONE RUBBINGS, Edmund V. Gillon, Jr. 43 photographs, 226 rubbings show heavily symbolic, macabre, sometimes humorous primitive American art. Up to early 19th century. 207pp. 8³/8 x 11¼.
21380-3 Pa. $4.00

L.J.M. DAGUERRE: THE HISTORY OF THE DIORAMA AND THE DAGUERREOTYPE, Helmut and Alison Gernsheim. Definitive account. Early history, life and work of Daguerre; discovery of daguerreotype process; diffusion abroad; other early photography. 124 illustrations. 226pp. 6¹/6 x 9¼. 22290-X Pa. $4.00

PHOTOGRAPHY AND THE AMERICAN SCENE, Robert Taft. The basic book on American photography as art, recording form, 1839-1889. Development, influence on society, great photographers, types (portraits, war, frontier, etc.), whatever else needed. Inexhaustible. Illustrated with 322 early photos, daguerreotypes, tintypes, stereo slides, etc. 546pp. 6¹/8 x 9¼. 21201-7 Pa. $5.95

PHOTOGRAPHIC SKETCHBOOK OF THE CIVIL WAR, Alexander Gardner. Reproduction of 1866 volume with 100 on-the-field photographs: Manassas, Lincoln on battlefield, slave pens, etc. Introduction by E.F. Bleiler. 224pp. 10¾ x 9.
22731-6 Pa. $6.00

THE MOVIES: A PICTURE QUIZ BOOK, Stanley Appelbaum & Hayward Cirker. Match stars with their movies, name actors and actresses, test your movie skill with 241 stills from 236 great movies, 1902-1959. Indexes of performers and films. 128pp. 8³/8 x 9¼. 20222-4 Pa. $2.50

THE TALKIES, Richard Griffith. Anthology of features, articles from Photoplay, 1928-1940, reproduced complete. Stars, famous movies, technical features, fabulous ads, etc.; Garbo, Chaplin, King Kong, Lubitsch, etc. 4 color plates, scores of illustrations. 327pp. 8³/8 x 11¼. 22762-6 Pa. $6.95

THE MOVIE MUSICAL FROM VITAPHONE TO "42ND STREET," edited by Miles Kreuger. Relive the rise of the movie musical as reported in the pages of Photoplay magazine (1926-1933): every movie review, cast list, ad, and record review; every significant feature article, production still, biography, forecast, and gossip story. Profusely illustrated. 367pp. 8³/8 x 11¼. 23154-2 Pa. $7.95

JOHANN SEBASTIAN BACH, Philipp Spitta. Great classic of biography, musical commentary, with hundreds of pieces analyzed. Also good for Bach's contemporaries. 450 musical examples. Total of 1799pp.
EUK 22278-0, 22279-9 Clothbd., Two vol. set $25.00

BEETHOVEN AND HIS NINE SYMPHONIES, Sir George Grove. Thorough history, analysis, commentary on symphonies and some related pieces. For either beginner or advanced student. 436 musical passages. 407pp. 20334-4 Pa. $4.00

MOZART AND HIS PIANO CONCERTOS, Cuthbert Girdlestone. The only full-length study. Detailed analyses of all 21 concertos, sources; 417 musical examples. 509pp. 21271-8 Pa. $6.00

THE FITZWILLIAM VIRGINAL BOOK, edited by J. Fuller Maitland, W.B. Squire. Famous early 17th century collection of keyboard music, 300 works by Morley, Byrd, Bull, Gibbons, etc. Modern notation. Total of 938pp. 8 3/8 x 11.
ECE 21068-5, 21069-3 Pa., Two vol. set $15.00

COMPLETE STRING QUARTETS, Wolfgang A. Mozart. Breitkopf and Härtel edition. All 23 string quartets plus alternate slow movement to K156. Study score. 277pp. 9 3/8 x 12 1/4.
22372-8 Pa. $6.00

COMPLETE SONG CYCLES, Franz Schubert. Complete piano, vocal music of Die Schöne Müllerin, Die Winterreise, Schwanengesang. Also Drinker English singing translations. Breitkopf and Härtel edition. 217pp. 9 3/8 x 12 1/4.
22649-2 Pa. $5.00

THE COMPLETE PRELUDES AND ETUDES FOR PIANOFORTE SOLO, Alexander Scriabin. All the preludes and etudes including many perfectly spun miniatures. Edited by K.N. Igumnov and Y.I. Mil'shteyn. 250pp. 9 x 12.
22919-X Pa. $6.00

TRISTAN UND ISOLDE, Richard Wagner. Full orchestral score with complete instrumentation. Do not confuse with piano reduction. Commentary by Felix Mottl, great Wagnerian conductor and scholar. Study score. 655pp. 8 1/8 x 11.
22915-7 Pa. $11.95

FAVORITE SONGS OF THE NINETIES, ed. Robert Fremont. Full reproduction, including covers, of 88 favorites: Ta-Ra-Ra-Boom-De-Aye, The Band Played On, Bird in a Gilded Cage, Under the Bamboo Tree, After the Ball, etc. 401pp. 9 x 12.
EBE 21536-9 Pa. $6.95

SOUSA'S GREAT MARCHES IN PIANO TRANSCRIPTION: ORIGINAL SHEET MUSIC OF 23 WORKS, John Philip Sousa. Selected by Lester S. Levy. Playing edition includes: The Stars and Stripes Forever, The Thunderer, The Gladiator, King Cotton, Washington Post, much more. 24 illustrations. 111pp. 9 x 12.
USO 23132-1 Pa. $3.50

CLASSIC PIANO RAGS, selected with an introduction by Rudi Blesh. Best ragtime music (1897-1922) by Scott Joplin, James Scott, Joseph F. Lamb, Tom Turpin, 9 others. Printed from best original sheet music, plus covers. 364pp. 9 x 12.
EBE 20469-3 Pa. $7.50

ANALYSIS OF CHINESE CHARACTERS, C.D. Wilder, J.H. Ingram. 1000 most important characters analyzed according to primitives, phonetics, historical development. Traditional method offers mnemonic aid to beginner, intermediate student of Chinese, Japanese. 365pp.
23045-7 Pa. $4.00

MODERN CHINESE: A BASIC COURSE, Faculty of Peking University. Self study, classroom course in modern Mandarin. Records contain phonetics, vocabulary, sentences, lessons. 249 page book contains all recorded text, translations, grammar, vocabulary, exercises. Best course on market. 3 12" 33 1/3 monaural records, book, album.
98832-5 Set $12.50

CONSTRUCTION OF AMERICAN FURNITURE TREASURES, Lester Margon. 344 detail drawings, complete text on constructing exact reproductions of 38 early American masterpieces: Hepplewhite sideboard, Duncan Phyfe drop-leaf table, mantel clock, gate-leg dining table, Pa. German cupboard, more. 38 plates. 54 photographs. 168pp. 8³/8 x 11¼. 23056-2 Pa. $4.00

JEWELRY MAKING AND DESIGN, Augustus F. Rose, Antonio Cirino. Professional secrets revealed in thorough, practical guide: tools, materials, processes; rings, brooches, chains, cast pieces, enamelling, setting stones, etc. Do not confuse with skimpy introductions: beginner can use, professional can learn from it. Over 200 illustrations. 306pp. 21750-7 Pa. $3.00

METALWORK AND ENAMELLING, Herbert Maryon. Generally conceded best all-around book. Countless trade secrets: materials, tools, soldering, filigree, setting, inlay, niello, repoussé, casting, polishing, etc. For beginner or expert. Author was foremost British expert. 330 illustrations. 335pp. 22702-2 Pa. $4.00

WEAVING WITH FOOT-POWER LOOMS, Edward F. Worst. Setting up a loom, beginning to weave, constructing equipment, using dyes, more, plus over 285 drafts of traditional patterns including Colonial and Swedish weaves. More than 200 other figures. For beginning and advanced. 275pp. 8¾ x 6³/8. 23064-3 Pa. $4.50

WEAVING A NAVAJO BLANKET, Gladys A. Reichard. Foremost anthropologist studied under Navajo women, reveals every step in process from wool, dyeing, spinning, setting up loom, designing, weaving. Much history, symbolism. With this book you could make one yourself. 97 illustrations. 222pp. 22992-0 Pa. $3.00

NATURAL DYES AND HOME DYEING, Rita J. Adrosko. Use natural ingredients: bark, flowers, leaves, lichens, insects etc. Over 135 specific recipes from historical sources for cotton, wool, other fabrics. Genuine premodern handicrafts. 12 illustrations. 160pp. 22688-3 Pa. $2.00

DRIED FLOWERS, Sarah Whitlock and Martha Rankin. Concise, clear, practical guide to dehydration, glycerinizing, pressing plant material, and more. Covers use of silica gel. 12 drawings. Originally titled "New Techniques with Dried Flowers." 32pp. 21802-3 Pa. $1.00

THOMAS NAST: CARTOONS AND ILLUSTRATIONS, with text by Thomas Nast St. Hill. Father of American political cartooning. Cartoons that destroyed Tweed Ring; inflation, free love, church and state; original Republican elephant and Democratic donkey; Santa Claus; more. 117 illustrations. 146pp. 9 x 12.
22983-1 Pa. $4.00
23067-8 Clothbd. $8.50

FREDERIC REMINGTON: 173 DRAWINGS AND ILLUSTRATIONS. Most famous of the Western artists, most responsible for our myths about the American West in its untamed days. Complete reprinting of *Drawings of Frederic Remington* (1897), plus other selections. 4 additional drawings in color on covers. 140pp. 9 x 12.
20714-5 Pa. $5.00

SLEEPING BEAUTY, illustrated by Arthur Rackham. Perhaps the fullest, most delightful version ever, told by C.S. Evans. Rackham's best work. 49 illustrations. 110pp. 7⅞ x 10¾. 22756-1 Pa. $2.00

THE WONDERFUL WIZARD OF OZ, L. Frank Baum. Facsimile in full color of America's finest children's classic. Introduction by Martin Gardner. 143 illustrations by W.W. Denslow. 267pp. 20691-2 Pa. $3.00

GOOPS AND HOW TO BE THEM, Gelett Burgess. Classic tongue-in-cheek masquerading as etiquette book. 87 verses, 170 cartoons as Goops demonstrate virtues of table manners, neatness, courtesy, more. 88pp. 6½ x 9¼. 22233-0 Pa. $2.00

THE BROWNIES, THEIR BOOK, Palmer Cox. Small as mice, cunning as foxes, exuberant, mischievous, Brownies go to zoo, toy shop, seashore, circus, more. 24 verse adventures. 266 illustrations. 144pp. 6⅝ x 9¼. 21265-3 Pa. $2.50

BILLY WHISKERS: THE AUTOBIOGRAPHY OF A GOAT, Frances Trego Montgomery. Escapades of that rambunctious goat. Favorite from turn of the century America. 24 illustrations. 259pp. 22345-0 Pa. $2.75

THE ROCKET BOOK, Peter Newell. Fritz, janitor's kid, sets off rocket in basement of apartment house; an ingenious hole punched through every page traces course of rocket. 22 duotone drawings, verses. 48pp. 6⅞ x 8⅜. 22044-3 Pa. $1.50

CUT AND COLOR PAPER MASKS, Michael Grater. Clowns, animals, funny faces . . . simply color them in, cut them out, and put them together, and you have 9 paper masks to play with and enjoy. Complete instructions. Assembled masks shown in full color on the covers. 32pp. 8¼ x 11. 23171-2 Pa. $1.50

THE TALE OF PETER RABBIT, Beatrix Potter. The inimitable Peter's terrifying adventure in Mr. McGregor's garden, with all 27 wonderful, full-color Potter illustrations. 55pp. 4¼ x 5½. USO 22827-4 Pa. $1.00

THE TALE OF MRS. TIGGY-WINKLE, Beatrix Potter. Your child will love this story about a very special hedgehog and all 27 wonderful, full-color Potter illustrations. 57pp. 4¼ x 5½. USO 20546-0 Pa. $1.00

THE TALE OF BENJAMIN BUNNY, Beatrix Potter. Peter Rabbit's cousin coaxes him back into Mr. McGregor's garden for a whole new set of adventures. A favorite with children. All 27 full-color illustrations. 59pp. 4¼ x 5½. USO 21102-9 Pa. $1.00

THE MERRY ADVENTURES OF ROBIN HOOD, Howard Pyle. Facsimile of original (1883) edition, finest modern version of English outlaw's adventures. 23 illustrations by Pyle. 296pp. 6½ x 9¼. 22043-5 Pa. $4.00

TWO LITTLE SAVAGES, Ernest Thompson Seton. Adventures of two boys who lived as Indians; explaining Indian ways, woodlore, pioneer methods. 293 illustrations. 286pp. 20985-7 Pa. $3.00

HOUDINI ON MAGIC, Harold Houdini. Edited by Walter Gibson, Morris N. Young. How he escaped; exposés of fake spiritualists; instructions for eye-catching tricks; other fascinating material by and about greatest magician. 155 illustrations. 280pp. 20384-0 Pa. $2.75

HANDBOOK OF THE NUTRITIONAL CONTENTS OF FOOD, U.S. Dept. of Agriculture. Largest, most detailed source of food nutrition information ever prepared. Two mammoth tables: one measuring nutrients in 100 grams of edible portion; the other, in edible portion of 1 pound as purchased. Originally titled Composition of Foods. 190pp. 9 x 12. 21342-0 Pa. $4.00

COMPLETE GUIDE TO HOME CANNING, PRESERVING AND FREEZING, U.S. Dept. of Agriculture. Seven basic manuals with full instructions for jams and jellies; pickles and relishes; canning fruits, vegetables, meat; freezing anything. Really good recipes, exact instructions for optimal results. Save a fortune in food. 156 illustrations. 214pp. 6⅛ x 9¼. 22911-4 Pa. $2.50

THE BREAD TRAY, Louis P. De Gouy. Nearly every bread the cook could buy or make: bread sticks of Italy, fruit breads of Greece, glazed rolls of Vienna, everything from corn pone to croissants. Over 500 recipes altogether. including buns, rolls, muffins, scones, and more. 463pp. 23000-7 Pa. $4.00

CREATIVE HAMBURGER COOKERY, Louis P. De Gouy. 182 unusual recipes for casseroles, meat loaves and hamburgers that turn inexpensive ground meat into memorable main dishes: Arizona chili burgers, burger tamale pie, burger stew, burger corn loaf, burger wine loaf, and more. 120pp. 23001-5 Pa. $1.75

LONG ISLAND SEAFOOD COOKBOOK, J. George Frederick and Jean Joyce. Probably the best American seafood cookbook. Hundreds of recipes. 40 gourmet sauces, 123 recipes using oysters alone! All varieties of fish and seafood amply represented. 324pp. 22677-8 Pa. $3.50

THE EPICUREAN: A COMPLETE TREATISE OF ANALYTICAL AND PRACTICAL STUDIES IN THE CULINARY ART, Charles Ranhofer. Great modern classic. 3,500 recipes from master chef of Delmonico's, turn-of-the-century America's best restaurant. Also explained, many techniques known only to professional chefs. 775 illustrations. 1183pp. 6⅝ x 10. 22680-8 Clothbd. $22.50

THE AMERICAN WINE COOK BOOK, Ted Hatch. Over 700 recipes: old favorites livened up with wine plus many more: Czech fish soup, quince soup, sauce Perigueux, shrimp shortcake, filets Stroganoff, cordon bleu goulash, jambonneau, wine fruit cake, more. 314pp. 22796-0 Pa. $2.50

DELICIOUS VEGETARIAN COOKING, Ivan Baker. Close to 500 delicious and varied recipes: soups, main course dishes (pea, bean, lentil, cheese, vegetable, pasta, and egg dishes), savories, stews, whole-wheat breads and cakes, more. 168pp. USO 22834-7 Pa. $2.00